# The Trigonometry Tryst

# The Trigonometry Tryst

• • •

*J. A. Bailey*

© 2017 J. A. Bailey
All rights reserved.

ISBN-13: 9781539695356
ISBN-10: 1539695352
Library of Congress Control Number: 2016917900
CreateSpace Independent Publishing Platform
North Charleston, South Carolina

*To my grandfather Pappa. Thank you.*

# Unit University

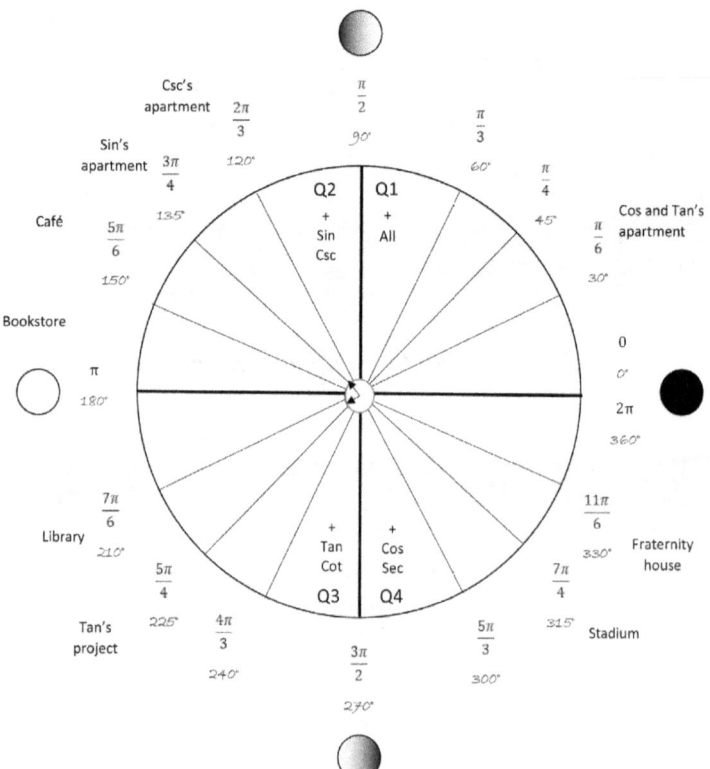

## Trigonometry Special Angles Table

| Degrees | Radians | $\sin\theta$ | $\cos\theta$ | $\tan\theta$ | $\cot\theta$ | $\sec\theta$ | $\csc\theta$ |
|---|---|---|---|---|---|---|---|
| 0° | 0 | 0 | 1 | 0 | undefined | 1 | undefined |
| 30° | $\dfrac{\pi}{6}$ | $\dfrac{1}{2}$ | $\dfrac{\sqrt{3}}{2}$ | $\dfrac{\sqrt{3}}{3}$ | $\sqrt{3}$ | $\dfrac{2\sqrt{3}}{3}$ | 2 |
| 45° | $\dfrac{\pi}{4}$ | $\dfrac{\sqrt{2}}{2}$ | $\dfrac{\sqrt{2}}{2}$ | 1 | 1 | $\sqrt{2}$ | $\sqrt{2}$ |
| 60° | $\dfrac{\pi}{3}$ | $\dfrac{\sqrt{3}}{2}$ | $\dfrac{1}{2}$ | $\sqrt{3}$ | $\dfrac{\sqrt{3}}{3}$ | 2 | $\dfrac{2\sqrt{3}}{3}$ |
| 90° | $\dfrac{\pi}{2}$ | 1 | 0 | undefined | 0 | undefined | 1 |
| 120° | $\dfrac{2\pi}{3}$ | $\dfrac{\sqrt{3}}{2}$ | $-\dfrac{1}{2}$ | $-\sqrt{3}$ | $-\dfrac{\sqrt{3}}{3}$ | $-2$ | $\dfrac{2\sqrt{3}}{3}$ |
| 135° | $\dfrac{3\pi}{4}$ | $\dfrac{\sqrt{2}}{2}$ | $-\dfrac{\sqrt{2}}{2}$ | $-1$ | $-1$ | $-\sqrt{2}$ | $\sqrt{2}$ |
| 150° | $\dfrac{5\pi}{6}$ | $\dfrac{1}{2}$ | $-\dfrac{\sqrt{3}}{2}$ | $-\dfrac{\sqrt{3}}{3}$ | $-\sqrt{3}$ | $-\dfrac{2\sqrt{3}}{3}$ | 2 |
| 180° | $\pi$ | 0 | $-1$ | 0 | undefined | $-1$ | undefined |
| 210° | $\dfrac{7\pi}{6}$ | $-\dfrac{1}{2}$ | $-\dfrac{\sqrt{3}}{2}$ | $\dfrac{\sqrt{3}}{3}$ | $\sqrt{3}$ | $-\dfrac{2\sqrt{3}}{3}$ | $-2$ |
| 225° | $\dfrac{5\pi}{4}$ | $-\dfrac{\sqrt{2}}{2}$ | $-\dfrac{\sqrt{2}}{2}$ | 1 | 1 | $-\sqrt{2}$ | $-\sqrt{2}$ |
| 240° | $\dfrac{4\pi}{3}$ | $-\dfrac{\sqrt{3}}{2}$ | $-\dfrac{1}{2}$ | $\sqrt{3}$ | $\dfrac{\sqrt{3}}{3}$ | $-2$ | $\dfrac{2\sqrt{3}}{3}$ |
| 270° | $\dfrac{3\pi}{2}$ | $-1$ | 0 | undefined | 0 | undefined | $-1$ |
| 300° | $\dfrac{5\pi}{3}$ | $-\dfrac{\sqrt{3}}{2}$ | $\dfrac{1}{2}$ | $-\sqrt{3}$ | $-\dfrac{\sqrt{3}}{3}$ | 2 | $-\dfrac{2\sqrt{3}}{3}$ |
| 315° | $\dfrac{7\pi}{4}$ | $-\dfrac{\sqrt{2}}{2}$ | $\dfrac{\sqrt{2}}{2}$ | $-1$ | $-1$ | $\sqrt{2}$ | $-\sqrt{2}$ |
| 330° | $\dfrac{11\pi}{6}$ | $-\dfrac{1}{2}$ | $\dfrac{\sqrt{3}}{2}$ | $-\dfrac{\sqrt{3}}{3}$ | $-\sqrt{3}$ | $\dfrac{2\sqrt{3}}{3}$ | $-2$ |
| 360° | $2\pi$ | 0 | 1 | 0 | undefined | 1 | undefined |

## Right-Triangle Definition

$0° < \theta < \dfrac{\pi}{2}$ or $0° < \theta < 90°$

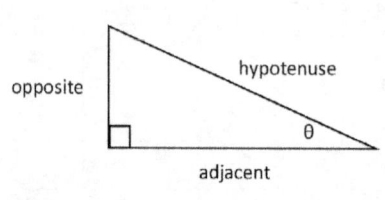

## Unit-Circle Definition

$\theta$ is any angle.

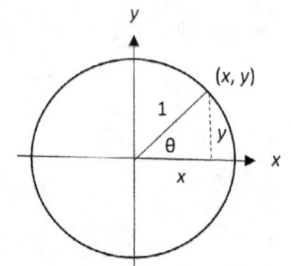

### Primary Functions

$\sin \theta = \dfrac{opposite}{hypotenuse}$

$\cos \theta = \dfrac{adjacent}{hypotenuse}$

$\tan \theta = \dfrac{opposite}{adjacent}$

### Secondary Functions

$\csc \theta = \dfrac{hypotenuse}{opposite}$

$\sec \theta = \dfrac{hypotenuse}{adjacent}$

$\cot \theta = \dfrac{adjacent}{opposite}$

### Primary Functions

$\sin \theta = \dfrac{y}{1} = y$

$\cos \theta = \dfrac{x}{1} = x$

$\tan \theta = \dfrac{y}{x}$

### Secondary Functions

$\csc \theta = \dfrac{1}{y}$

$\sec \theta = \dfrac{1}{x}$

$\cot \theta = \dfrac{x}{y}$

# Trigonometry Formulas and Identities

## Reciprocal Identities

$\sin\theta = \dfrac{1}{\csc\theta} \qquad \csc\theta = \dfrac{1}{\sin\theta}$

$\cos\theta = \dfrac{1}{\sec\theta} \qquad \sec\theta = \dfrac{1}{\cos\theta}$

$\tan\theta = \dfrac{1}{\cot\theta} \qquad \cot\theta = \dfrac{1}{\tan\theta}$

## Pythagorean Identities

$\sin^2\theta + \cos^2\theta = 1$

$\tan^2\theta + 1 = \sec^2\theta$

$1 + \cot^2\theta = \csc^2\theta$

## Even/Odd Formulas

$\sin(-\theta) = -\sin\theta \, (odd)$

$\csc(-\theta) = -\csc\theta \, (odd)$

$\cos(-\theta) = \cos\theta \, (even)$

$\sec(-\theta) = \sec\theta \, (even)$

$\tan(-\theta) = -\tan\theta \, (odd)$

$\cot(-\theta) = -\cot\theta \, (odd)$

## Periodic Formulas
If *n* is an integer

$\sin(\theta + 2\pi n) = \sin\theta$

$\cos(\theta + 2\pi n) = \cos\theta$

$\tan(\theta + \pi n) = \tan\theta$

$\csc(\theta + 2\pi n) = \csc\theta$

$\sec(\theta + 2\pi n) = \sec\theta$

$\cot(\theta + \pi n) = \cot\theta$

## Quotient Identities

$\tan\theta = \dfrac{\sin\theta}{\cos\theta} \qquad \cot\theta = \dfrac{\cos\theta}{\sin\theta}$

## Cofunction Identities

$\sin\left[\dfrac{\pi}{2} - \theta\right] = \cos\theta \qquad \cos\left[\dfrac{\pi}{2} - \theta\right] = \sin\theta$

$\tan\left[\dfrac{\pi}{2} - \theta\right] = \cot\theta \qquad \cot\left[\dfrac{\pi}{2} - \theta\right] = \tan\theta$

$\csc\left[\dfrac{\pi}{2} - \theta\right] = \sec\theta \qquad \sec\left[\dfrac{\pi}{2} - \theta\right] = \csc\theta$

## Degrees-to-Radians Conversion

$\theta° * \dfrac{\pi \, rad}{180°}$

## Radians-to-Degrees Conversion

$rad * \dfrac{180°}{\pi \, rad}$

## Inverse Trigonometric Functions

If $\sin\theta = x$,
$\theta = \arcsin(x) = \sin^{-1}(x)$
"The angle whose sin is $x$."

If $\cos\theta = x$,
$\theta = \arccos(x) = \cos^{-1}(x)$
"The angle whose cos is $x$."

If $\tan\theta = x$,
$\theta = \arctan(x) = \tan^{-1}(x)$
"The angle whose tan is $x$."

If $\csc\theta = x$,
$\theta = \text{arccsc}(x) = \csc^{-1}(x)$
"The angle whose csc is $x$."

If $\sec\theta = x$,
$\theta = \text{arcsec}(x) = \sec^{-1}(x)$
"The angle whose sec is $x$."

If $\cot\theta = x$,
$\theta = \text{arccot}(x) = \cot^{-1}(x)$
"The angle whose cot is $x$."

## Double-Angle Formulas

$\sin(2\theta) = 2\sin\theta\cos\theta$

$\cos(2\theta) = \cos^2\theta - \sin^2\theta$

$\qquad = 2\cos^2\theta - 1$

$\qquad = 1 - 2\sin^2\theta$

$\tan(2\theta) = \dfrac{2\tan\theta}{1 - \tan^2\theta}$

## Half-Angle Formulas

$\sin^2\theta = \dfrac{1}{2}(1 - \cos(2\theta))$

$\cos^2\theta = \dfrac{1}{2}(1 + \cos(2\theta))$

$\tan^2\theta = \dfrac{1-\cos(2\theta)}{1+\cos(2\theta)}$

## Sum and Difference Formulas

$\sin(\alpha \pm \beta) = \sin\alpha\cos\beta \pm \cos\alpha\sin\beta$

$\cos(\alpha \pm \beta) = \cos\alpha\cos\beta \mp \sin\alpha\sin\beta$

$\tan(\alpha \pm \beta) = \dfrac{\tan\alpha \pm \tan\beta}{1 \mp \tan\alpha\tan\beta}$

## Angular and Linear Velocity

Angular Velocity: $\omega = \dfrac{\theta}{t}$

Linear Velocity: $V = r\omega$

## Product-to-Sum Formulas

$\sin\alpha\sin\beta = \dfrac{1}{2}[\cos(\alpha - \beta) - \cos(\alpha + \beta)]$

$\cos\alpha\cos\beta = \dfrac{1}{2}[\cos(\alpha - \beta) + \cos(\alpha + \beta)]$

$\sin\alpha\cos\beta = \dfrac{1}{2}[\sin(\alpha + \beta) + \sin(\alpha - \beta)]$

$\cos\alpha\sin\beta = \dfrac{1}{2}[\sin(\alpha + \beta) - \sin(\alpha - \beta)]$

## Sum-to-Product Formulas

$\sin\alpha + \sin\beta = 2\sin\left(\dfrac{\alpha+\beta}{2}\right)\cos\left(\dfrac{\alpha-\beta}{2}\right)$

$\sin\alpha - \sin\beta = 2\cos\left(\dfrac{\alpha+\beta}{2}\right)\sin\left(\dfrac{\alpha-\beta}{2}\right)$

$\cos\alpha + \cos\beta = 2\cos\left(\dfrac{\alpha+\beta}{2}\right)\cos\left(\dfrac{\alpha-\beta}{2}\right)$

$\cos\alpha - \cos\beta = -2\sin\left(\dfrac{\alpha+\beta}{2}\right)\sin\left(\dfrac{\alpha-\beta}{2}\right)$

## Law of Sines

$\dfrac{\sin A}{a} = \dfrac{\sin B}{b} = \dfrac{\sin C}{c}$

## Law of Cosines

$c^2 = a^2 + b^2 - 2ab\cos C$

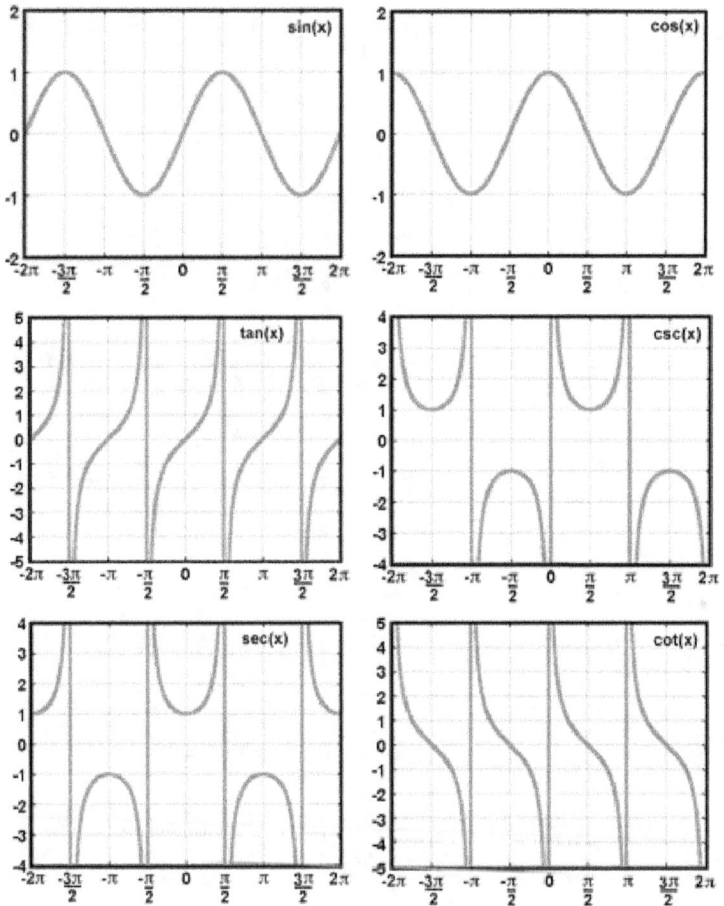

The scientific principles that man employs to obtain the foreknowledge of an eclipse, or of anything else relating to the motion of the heavenly bodies, are contained chiefly in that part of science that is called trigonometry, or the properties of a triangle, which, when applied to the study of the heavenly bodies, is called astronomy; when applied to direct the course of a ship on the ocean, it is called navigation; when applied to the construction of figures drawn by a ruler and compass, it is called geometry; when applied to the construction of plans of edifices, it is called architecture; when applied to the measurement of any portion of the surface of the earth, it is called land surveying. In fine, it is the soul of science. It is an eternal truth: it contains the mathematical demonstration of which man speaks, and the extent of its uses are unknown.

—Thomas Paine

# Author's Note
● ● ●

MATH IS ABOUT RELATIONSHIPS. TRIGONOMETRY is a branch of mathematics about relationships between lengths and angles of triangles. There are six trigonometric functions—sine (sin), cosine (cos), tangent (tan), cosecant (csc), secant (sec), and cotangent (cot), the only named characters in this book—defined as ratios between two sides of a right triangle. This book humanizes these relationships by creating protagonists out of the primary functions and antagonists out of the secondary functions. Words a student of trigonometry will recognize, such as "positive," "negative," and "undefined," will come alive as human feelings. Does the graph of each trigonometric function conform to that function's personality? Cofunctions, "complementary" relationships, quadrants, formulas, and identities will help determine which functions will be present at the Trigonometry Tryst.

The unit circle, a circle with a radius of one centered at the origin of the Cartesian coordinate system, is a place that defines the trigonometric functions and is used as a university setting for this novel. Note how the trigonometry characters move about the university on the unit circle, and try to determine where on the circumference of the university the tryst will be held and who will meet with whom. Enjoy this trigonometry journey!

CHAPTER 1

● ● ●

"Cos! Where are you headed?"

Cos looked up as she left the university bookstore with a bag holding two rolls of packing tape dangling at her side.

"Oh, hi, Csc. I'm helping Tan move. We ran out of packing tape. I didn't think it would take this much. She doesn't have a lot of stuff, but we're almost done."

A bitter-cold February wind was blowing so hard that the two girls' eyes watered. Cos held her scarf over her mouth to minimize the bite of the freezing air in her lungs as she breathed.

Csc seemed anxious. There were only a few months left in the second semester of their senior year at Unit

University, but Tan had informed her she was leaving the apartment she shared with Csc in the Q2 district and going to room with Cos in the Q1 district. Csc's nervous giggle was normal, but today her excited hazel eyes were also bloodshot.

"Oh yeah, I forgot you were coming to help Tan pack up," Csc said, talking quickly in little short breaths that had nothing to do with the cold air. "I'm sure she'll be happier over in the Q1 district rooming with you, even though she'll need to walk farther to class. Most of her classes are in the Q3 district…"

Csc's voice trailed off as she put her hands in her pockets and looked down at her shoes. One of her natural-blond curls hanging out from her beanie blew across her eyes. She bit one side of her bottom lip, brushed her hair away, and quickly looked back up at Cos, not knowing what to say next.

Cos looked at her for a couple of seconds. "Csc? Is everything OK? I'm sorry the roommate situation didn't work out with you and Tan. Tan just doesn't like this district. She doesn't like feeling negative all the time and wants to move to a district where the environment is more positive for her. To be honest, I don't like the Q2 district either. I have to work hard not to sink into a bad mood when I come over here."

"Well, it's worked out for me so far," Csc said. "Everyone loves the Q1 district, but I could never afford to live there. I don't know how the students that live there can do it, unless their parents are paying for it all."

"It's not th—" Cos got distracted when she saw Sin moving down the arc of sidewalk on the circumference of the campus. He was heading into the café next to the bookstore with his backpack slung over his shoulder. Csc turned her head to see what had caught Cos's attention.

"Sin likes this district well enough," Csc said with her head still turned over her shoulder, looking his way. "He likes that café, at least. Best hot spot in Q2. He spends hours in there studying, and he even lives in this district," Csc said, turning her head back around with a quiet, nervous giggle and rolling her feet out to stand on the sides of her shoes.

"He does?" Cos asked, surprised.

"Yep, right over in the apartment building at $3\pi/4$."

"Well, I'd better get back to helping Tan. Besides, my nose is about to freeze. I'll see you soon again, I hope," Cos said. She left feeling bad that Csc was on her own, needing a roommate with only a few more months in the semester. It wasn't likely that Csc would find one so close to the end of the academic year.

Her thoughts turned to Sin. Cos saw Sin daily on campus. Their paths crossed frequently, especially at $\pi/4$ in the Q1 district and $5\pi/4$ in the Q3 district, but they had never had a real conversation. When they passed each other, Sin always seemed to be thinking seriously about something and was too focused to look up and acknowledge anyone around him. He'd have his hands in the pockets of his brown leather jacket with his backpack casually hanging from his shoulder. His dark-brown hair matched his eyes. Even when the air was still, Sin walked as if he were alone on a beach, with dark clouds in the sky, the air chilly and blustery. He kept his head tucked into his jacket to protect himself from the wind. Despite his monochrome appearance, he stood out to Cos as one of the most distinct guys on campus.

As Cos made her way clockwise back to Tan and Csc's apartment, located at $2\pi/3$, she passed Sin's apartment building and wondered, *Why in the world would he live in this district?*

Tan had everything in boxes or bags when Cos walked into the apartment. "It must be windy out there this morning." Tan laughed when she saw Cos's short light-brown hair all askew.

"I just saw Csc," Cos said, smiling while she smoothed her hair after rubbing her hands together to warm them up. "She seems nervous that you're moving out." Cos took the rolls of packing tape out of the plastic bag and began taping up the boxes. "Have you even seen her this morning?"

"No, I haven't. Ouch!" A strand of Tan's long, smooth dark hair got caught on the packing tape as she taped a box of books closed. Tan wasn't about to give up any of her books, as reading was her favorite pastime. She quickly threw her hair up in a clip and stacked the box with the others waiting to be moved.

"Did you get to know her at all?" asked Cos as she tore a piece of transparent packing tape along the sharp, serrated metal edge. She looked over at Tan with caring blue eyes.

"A little. We don't have a lot in common. She was usually nervous, as if she was going to say the wrong thing. It's obvious she is unsure of herself but wants to make friends badly. I don't mind being alone and doing my own thing. She does mind it. I can see why she's nervous about my moving out. When I told her, she thought

*The Trigonometry Tryst*

I was moving out because I didn't like her. I had to convince her I was moving because I don't like the Q2 district. I dread coming into this district. I feel so negative here, and I just didn't think I could spend the last three months here when there was a perfect opportunity to move in with you in the Q1 district. Well, you know. You feel the same way. I didn't know it would get bad enough that I would need to move before graduating," Tan said in her usual matter-of-fact style. Her green eyes were bright but usually serious.

"Csc says she likes it," Cos said. "But I'm sure she would love to be in the Q1 district. Everyone loves it there. I was lucky to get the apartment when I did. She may not find a roommate this close to the end of the term, though."

"If only she didn't feel so secondary. I wonder what would give her more confidence in herself," Tan said.

Feeling secondary was something no one could accuse Tan of. Tan was graduating this year with a degree in architecture. She loved life, but a good life to her was not an evening of partying. She would rather have read a book or gone to dinner with a friend and talked about ideas. She was fine with or without people, but she preferred people who were responsible, edifying, and goal oriented. Just hanging out was painful, draining her creative energy and wasting her time.

Tan went on a date now and then but never felt serious about anyone, though she cared about people in general. Cos wished she could feel that sure of herself, and she was worried about graduating this year without meeting someone she was serious about. She knew opportunities would be leaner after she moved away from campus.

That reminded her: "Tan, do you mind if I pull up the Internet on your tablet since you haven't packed it yet? I want to check Unit Match."

Tan laughed. "Are you serious? I didn't know you had a profile on there."

*The Trigonometry Tryst*

Cos raised her eyebrows a few times, knowing that Tan was not making fun of her. She felt she could say anything to Tan, even if they were as different as they were. "Well, I do. You never know."

Cos pulled up Unit Match, which was Unit University's own online matchmaking social website. The students thought it was odd at first when the website had gone live about a year ago. Online matchmaking sites used to be for people who couldn't find anyone in their own communities and needed to check out opportunities outside. It was true, though, that some couples found each other online, never knowing they lived only a couple of streets apart. Unit Match soon became all the rage with the students. They were curious to see whether there was anyone interesting on campus whom they hadn't met. It also didn't hurt to see those with profiles on Unit Match whom they did know on campus and who were searching for potential relationships.

Cos logged on to her account and looked through the usual profiles that had been there for a while. There were a few new ones she skimmed through, but nothing caught her eye. There was one with no picture, which happened occasionally. Not posting a picture wasn't a good thing to do, because students assumed the person didn't like the way he or she looked or it was a smug statement declaring that looks didn't matter. Cos wasn't looking for statements. She was looking for connection. She knew it would be harder to find online, but she was willing to try anyway. For fun, she pulled up the profile without a picture, thinking it would be the usual. It wasn't:

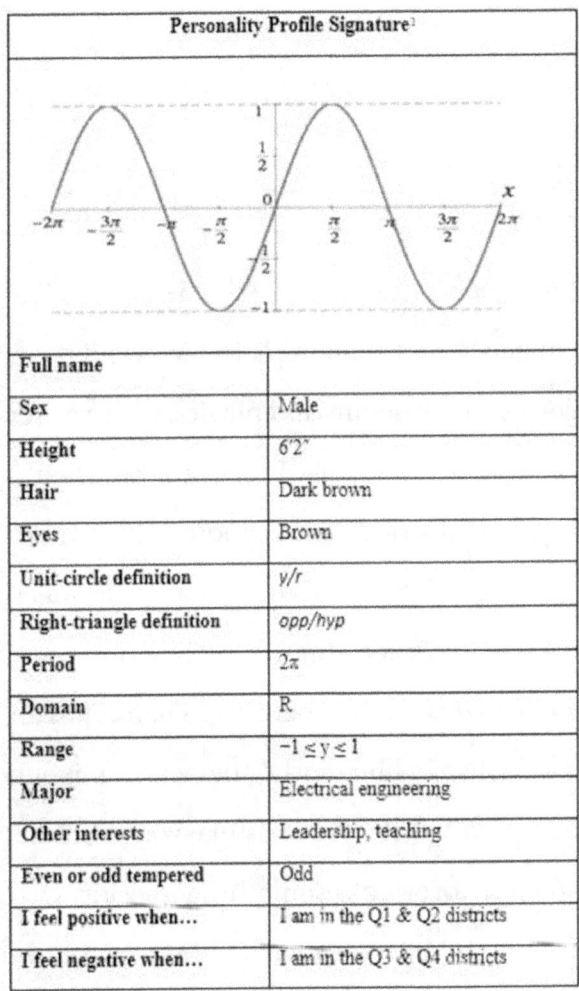

| **Personality Profile Signature**[1] | |
|---|---|
| Full name | |
| Sex | Male |
| Height | 6'2" |
| Hair | Dark brown |
| Eyes | Brown |
| Unit-circle definition | y/r |
| Right-triangle definition | opp/hyp |
| Period | $2\pi$ |
| Domain | R |
| Range | $-1 \leq y \leq 1$ |
| Major | Electrical engineering |
| Other interests | Leadership, teaching |
| Even or odd tempered | Odd |
| I feel positive when… | I am in the Q1 & Q2 districts |
| I feel negative when… | I am in the Q3 & Q4 districts |

---

[1] All Personality Profile graphs in this novel were obtained from "Trigonometric Function Graphs F($\pi$)," Edward Furey, 2017, HYPERLINK "http://www.calculatorsoup.com/" http://www.calculatorsoup.com (Online Calculator Resource).

| Invitation to Meet | |
|---|---|
| Date | April 30, 2017 |
| Time | 8:07 p.m. |
| Place | |

"Hmm…" Cos mused. "Here's someone that is almost a perfect match. Other than his definitions and the fact that I'm majoring in astronomy and physics, we have everything else in common. Oh, wait—I feel positive when I am in the Q1 and Q4 districts. That's another difference. That's strange, though. He posted an Invitation to Meet with a date and time, but not a meeting place. How would you know where to meet him? There isn't a name, just an e-mail address without any hint of who the person might be."

Tan scoffed lightly. "Sounds like you shouldn't waste your time if he overlooks something as critical as that in an Invitation to Meet. I guess he'll figure it out when no one shows up!"

Cos sighed and logged out of her profile. "OK, let's get this stuff moved."

CHAPTER 2

• • •

SEC WATCHED COS AND CSC when they turned to look at Sin going into the café. When they finally went their separate ways, Sec waited for Csc to walk in his direction. Maybe he could get some information about Cos from her. While Sec and Csc were not in the same social circles, Sec knew of Csc's reputation. He would never have admitted he intuitively recognized her weakness and was drawn to it. Somehow, he related to her flaws, but his social reputation told him he didn't have any himself. He wasn't going to be embarrassed striking up conversations with Csc. A star athlete befriending a socially awkward girl would bring him accolades.

Sec had a class with Cos and had tried talking with her a few times, but although she was polite, she wasn't open with him or as friendly as he had seen her be with other classmates. As a star athlete, he imagined she was intimidated by his prowess and popularity on the basketball court. However, when he saw she was friendly to other big-name athletes on campus, he began to feel secondary to her and angry that she would relate to him this way. His jealous nature made him determined to make her recognize him. Sec refused to consider that she sensed inferiority in him, to protect himself from feeling spurned and rejected.

About to pass by Sec, Csc was looking down at her feet as she walked, still feeling stung by the roommate situation and hoping Cos hadn't noticed. Suddenly, she felt someone pull hard on her sleeve, which almost made her lose her balance. Stunned, she looked up to see Sec.

"Hi, Csc! What are you up to?"

Csc smiled immediately, taking the bait that someone might be interested in being her friend. "Nothing in particular. Why?"

"No reason. Hey, would you want to get a bagel or something to drink at the café? I need a break from studying." He felt smooth as he detected a blush on Csc's face and felt more confident that she could be easily manipulated.

"Sure," Csc said bashfully. She couldn't believe that someone, especially Sec, actually wanted to spend time with her.

Csc followed Sec into the café, glancing around to make note of where Sin was sitting. Sin looked up and acknowledged them before promptly returning to what he was doing.

Sec waited for their orders and told Csc to go hold a table for them. The café was getting more crowded, with only a few tables to choose from. Csc chose a table not too near Sin, but she was very much aware of his presence. Her overall insecurities were usually magnified when she was near Sin. She always felt secondary to him and never initiated any conversation. Sin projected quality and substance, which made Csc feel inferior. She didn't have any reason to dislike him, but her mind-set ensured that Csc never approached with confidence anyone who she thought was beyond her, no matter how much she wanted to be in a better place herself.

It didn't take too long for Sec to bring their orders to the table. He set his toasted bagel and cream cheese with Csc's steaming-hot muffin and their drinks in front of them, and while returning the food tray, he subtly glanced over in Sin's direction. It turned out he didn't need to be so shrewd, as Sin seemed oblivious to his being there, engrossed in what he was doing on his laptop.

When Sec finally took his seat at the table, Csc began feeling anxious. She didn't mean to let out her signature nervous giggle, giving herself away, but it felt so automatic when she was in social situations. Sec appeared not to notice and instead took a big bite of his bagel; however, they both were aware of other students in the café looking in their direction. No one had ever expected to see the two of them eating together.

When he had swallowed enough to talk, he asked, "Do you know him?" and nodded in Sin's direction without looking over at him.

"Not really. I see him around campus, and he's in one of my classes. Why?" Csc asked.

"Just curious." Sec told himself to be more patient and make this meeting about Csc. He was going to need her collaboration if he was going to use her to his advantage. The sooner he had her wrapped around his little finger, the better.

"So I noticed you talking to Cos," Sec said. "She's lucky to be associated with a girl like you." While he took a sip of his drink, he observed her carefully over the rim of his cup.

Csc couldn't believe what she had just heard. She was naïve enough to take the flattery to heart. Maybe deep down she recognized the insincerity of it, but she was so starved for attention to fill up her emptiness that she let this tendency take the upper hand. Sec recognized the eager look in her eyes and knew this was going to be easier than he had thought.

"What are you majoring in?" he asked.

"Computer engineering with a minor in information technology."

"Cool. I figured you were smart." While delivering this bit of flattery, Sec's gears were turning in his mind. He was making a mental note of this information, knowing it would come in handy.

"How about you?" Csc asked to try to cover up the awkwardness she was feeling even though she was eating up the compliments.

"Marketing, with a minor in physics," Sec said with some pride. "Basketball takes a lot of my time," he added in a tone that reminded her who he was.

"Sweet. Well, I guess our majors don't exactly make us the perfect fit for study partners," Csc commented with a little disappointment showing. She wondered whether they would get together again if they didn't have a reason to.

"Sure they do. How does the saying go?" Sec scratched his head, pretending a memory lapse. "Something about the benefits of complementary relationships."

"I can't help you. I haven't heard that one before."

"Well, all it means is that I think we ought to be study partners."

Csc did her best to hide her excitement, but the sudden color in her cheeks gave her away. "Great," she said a little too quickly.

"You barely ate your muffin," Sec said, looking down at the crumbs surrounding the oversized muffin, as though she had mostly been picking at it.

"I'll just take it home with me. It will be a quick breakfast before class tomorrow. I'll get a bag on the way out."

"While you're getting the bag, you wouldn't mind doing me a favor, would you?" Sec normally would have waited a little longer before asking for favors, but he thought Csc was just too easy. He saw no reason to waste time, especially since the opportunity was ripe.

"Sure. Anything," Csc said. She would have done anything for Sec's approval.

CHAPTER 3

• • •

Sin had looked up from his laptop when Sec and Csc walked in the café earlier. Sec had spotted Sin instantly and nodded when they made eye contact. Sin had acknowledged him and Csc and then turned his focus back to his computer, knowing that neither of them would come over to his table and try to make conversation. He didn't know much about Csc. He had a class with her and felt she was smart but somewhat peculiar, in an unstable way. He wasn't sure what to make of her, as she usually avoided him intentionally when they passed each other on campus.

Too focused to give any more thought to Sec and Csc, Sin began to concentrate again. He was working on his personality profile for Unit Match. He had uploaded the basics a few days earlier, but he wanted to review it. Sin had lately been feeling he would like to meet some like-minded people he could spend time with. Even though he was surrounded by people at the university, they were only acquaintances.

In his estimation, most people were average. Conversation about anything interesting or exciting to him was virtually nonexistent. He enjoyed being social, but most of what everyone else considered to be social bored him. Not in an arrogant way—he just had goals and better things to do than to get drunk or hang out for hours without a purpose. He had worked hard all his life and had things he wanted to accomplish.

Over the last three years Sin had been at the university, a couple of clubs had interested him, but when he joined, he was disappointed to find that the members didn't take them too seriously. Meetings were often canceled, and the lack of functioning made him want to move on to other things. He was driven in a way that was hard to talk with other people about. Sin wasn't sure how to talk about wanting to be successful in his career, finances, health, family, and relationships, the whole person really, without people perceiving him to be self-righteous somehow. He was always surprised at this reaction from others, so he pulled back from sharing himself most of the time to avoid this, although his passion did not dissipate.

Sin knew himself well enough to know that deep down he dreamed of having friends and associates who felt the same way he did. He wanted friends who could share ideas, who wanted more out of life than mediocrity, and who knew how to create successful lives. *Where are these people? How do you find them?* He would ask himself these questions when he was discouraged. It was hard for him to settle for anything less. He could sink into a despondent mood now and then when dwelling on this lack in his life.

Sin had just entered in his profile that he was majoring in electrical engineering. He clicked the save button and decided to continue with his profile later. He had a short online open-book quiz to take on alternating currents, which he thought he would complete before he left the café.

Just finishing the sixth problem on the quiz, Sin heard a simultaneous groan come from other students who were also working on their computers in the café. He heard the same groan come out of his own mouth when he realized the Wi-Fi connection had gone down and he had lost the screen containing all his work on the quiz. He would need to start all over, as he had not submitted the work he had done so far, nor had he saved it. He kicked himself because he knew better. He fell back into his chair and looked up to see other frustrated patrons looking hopelessly at their computers. He also noticed that Sec and Csc were gone.

CHAPTER 4

● ● ●

Csc came back to her apartment and found that Tan and Cos had already left with the last of Tan's things. Tan's key was on the kitchen counter along with a note:

Csc,
Sorry I didn't see you before we left. I guess you had an early morning. Hope to see you on campus. Let me know if you need anything. Thanks for being my roommate. Come see Cos and me anytime!
Tan

Csc carried the note to the living room and slumped down on the secondhand sofa with tears stinging her eyes. The apartment seemed so quiet and empty. Normally, the quiet was because she and Tan were usually not home at the same time, but now it seemed even quieter, with her knowing that Tan was gone and she was alone again. She was always alone. The truth was that she had avoided Tan on purpose this morning. She liked Tan and wanted to be friends with her, but she knew Tan was better than she was. Tan didn't care what people thought. She had so much confidence and was one of the most anchored people Csc knew, but Csc couldn't figure out how to be that way herself. She just assumed Tan didn't like her and wouldn't have wanted her around this morning.

Csc was tired of being left behind. She wished she knew who she was, but even when she tried to act with confidence and self-assurance, people somehow picked up on her insecurity and never thought she was worth getting to know further. *What's wrong with me?* Her eyes overflowed with tears that fell down her cheeks.

Even as she was wiping away her tears, she wondered why she felt this way after her time with Sec today. She had a new friend now. She would meet with Sec again. It didn't matter that the favor he had asked of her at the café was a little strange. She felt part of something. Included. This wasn't a feeling she was used to, and she would do anything to hang on to it.

An emotional mess, Csc curled up on the sofa, pulled a throw blanket over herself, and cried until she fell asleep.

CHAPTER 5

● ● ●

UNIT UNIVERSITY'S CAMPUS WAS LAID out in a perfect circle with a one-unit radius. This unit circle was the basis for all the business conducted on campus. The layout of the circle determined the address system, the movements and directions of students, and how they related to one another. Students expressed themselves and interacted with one another in a variety of ways depending on where they were located on the unit circle. The campus's circumference was studded with student apartment housing, shops, and services catering to student life.

Addresses were easy to figure out, since all were measured against the main entrance at 0°, or standard position. Any address was measured from this point counterclockwise around the circle. Address numbers could be in radians or degrees but were marked on buildings in radians. If an address on a package or an envelope was written in degree mode instead of radian mode, the post office would just use the following conversion to change from degrees to radians for delivery:

$$\theta° * \frac{\pi\ rad}{180°}$$

A clock tower stood at the center of campus, at the origin (0, 0) of the circle. It stood tall, with four clock faces resting in a cube that sat on a square red-brick column. A tetrahedron enjoyed the pinnacle of the campus; it was positioned at the top of the clock tower with its square base joining perfectly to the clock-faced cube below it. Its apex could be seen from any point on the circumference of the campus's unit circle, making it a central reference point.

*J. A. Bailey*

While the base of the clock tower was centered on the origin of the campus, (0, 0), it was also centered in a perfect circle of smooth white cement. This circle was divided into unequal sectors depicting special angles such as 30°, 45°, and 60°, measured from the origin and standard position engraved in the cement. The special angles engraved in each quadrant of the circle of cement acted as a map and compass for students. They could walk along the radius in the direction indicated by their chosen angle, and they would then end up at the desired address on the circumference of the campus. At each of the quadrantal angles (0°, 90°, 180°, 270°) on campus stood striking black iron double gates hinged to high red-brick walls circumscribing the campus. These four gates were the only entrances to the campus, and students could feel their commanding presence as they crossed the threshold into a world where they learned how to think, work, create, and relate to other people.

Each gate entrance, though designed in a beautiful black iron motif with intricate patterns, shapes, and figures, was distinguished by its own unique circular insignia, which rested at the crest of a black iron arc spanning the width of the entrance. Students walked by or through these gates and under these symbols multiple times a day, eventually taking the symbolic patterns for granted.

Each insignia was circumscribed by the words "Unit University." Inside this periphery of letters was positioned an inner circle, which at the gate at the quadrantal angle of $0°$ was empty of any further design. Only a smooth, circular face of black iron filled the metallic ring of words, serving as the circular constraint.

If students walked through this one gate only, they might not recognize the insignia as anything more than a simple unadorned circle, absent any symbolic meaning. However, if students walked through the gate at the $90°$ quadrantal angle of the campus, they might look up to see that the circle suspended over that gate was divided vertically down the middle: the left half was filled with black iron while the right half was filled with a silvery metallic sheen that shimmered on sunny days. At the quadrantal angle of 180°, the entire circular space was filled with this metallic luster.

By the time students walked through the quadrantal-angle gate at 270°, if they were familiar with all four gates and the differences in the symbols suspended over each gate, they might have made a connection. At the 270° gate, they would have noticed that the circular space was divided in half again, but with the opposite design arrangement of that at the 90° gate. This time, the left half of the circular space was filled with the silvery metallic patina, leaving the right half to retain the opposite lackluster of the black iron.

The pattern of insignias connecting the four gates represented the phases of the moon: a new moon at the 0° gate, a first-quarter moon at the 90° gate, a full moon at the 180° gate, a third-quarter moon at the 270° gate, and back to a new moon again at 0°. The recurring, periodic cycle started anew each month.

As Cos was an astronomy major, she couldn't help noticing this pattern in the gates' insignias. She kept a calendar of the phases of the moon on her bedroom wall and was always keenly aware of the moon's position and illuminated phase at any time. Cos loved to look at the moon when she was walking on campus after dark, but she also enjoyed its beauty when she could see the moon during the day.

CHAPTER 6

● ● ●

SIN STEPPED OFF THE UNIVERSITY bus, which made continuous revolutions around the university, picking up and dropping off students and faculty, providing a faster alternative to walking when needed. The unmanned electric bus conveniently kept to its defined concentric circle just outside the university gates.

Traveling at forty miles per hour, the bus was programmed to stop at each of the four university gates to unload and load passengers. Sin liked to walk but enjoyed the ride during inclement weather. A storm was just subsiding, but it looked as if it would start again at any minute.

When he was dropped off outside the university gate at 180°, Sin set his backpack down on a stone bench and took out his circuits textbook. He wanted to finish reading a chapter before his class started. Opening his book, he caught sight of a large puddle of water created by the rainstorm. The wind started to blow hard again. He watched tiny ripples in the puddle created by the wind. They reminded him of alternating currents, which he studied daily as part of his electrical engineering major. He knew that the most common wave form of alternating current was the sine wave:

Watching one of the ripples dissipate, Sin thought about the Invitation to Meet he had posted on Unit Match. He smiled at the thought of students seeing his Invitation, baffled by the missing information: his name and where to meet. He had left this key information out intentionally, knowing if anyone understood the rest of the information in the Invitation and in his profile, and if the person was interested, he or she would be able to fill in the missing piece of the meeting location, and the meeting would take place. This strategy would filter out students who didn't have an understanding, so developing a connection with them wouldn't be likely anyway. He had gotten this idea from a story he once read called "The Telegraph," which went something like this:

> Back when the telegraph was the fastest means of long-distance communication, there was a story about a young man who applied for a job as a Morse-code operator. Answering an ad in the newspaper, he went to the address that was listed. When he arrived, he entered a large, noisy office. In the background a telegraph clacked away. A sign on the receptionist's counter instructed job applicants to fill out a form and wait until they were summoned to enter the inner office.

The young man completed his form and sat down with seven other waiting applicants. After a few minutes, the young man stood up, crossed the room to the door of the inner office, and walked right in. Naturally, the other applicants perked up, wondering what was going on. Why had this man been so bold? They muttered among themselves that they hadn't heard any summons yet. They took more than a little satisfaction in assuming the young man who went into the office would be reprimanded for his presumption and disqualified for the job.

Within a few minutes the young man emerged from the inner office escorted by the interviewer, who announced to the other applicants, "Gentlemen, thank you very much for coming, but the job has been filled by this young man."

The other applicants began grumbling to each other, and one spoke up, saying, "Wait a minute; I don't understand. He was the last to come in, and we never even got a chance to be interviewed. Yet he got the job. That's not fair!"

The employer responded, "All the time you've been sitting here, the telegraph has been ticking out the following message in Morse code: 'If you understand this message, then come right in. The job is yours.' None of you heard it or understood it. This young man did. The job is his."[2]

Sin could be patient, even when looking forward to seeing whether anyone was listening and interested in his own "Morse code." He knew that someone who was "tuned in" would have a Morse code he would be interested in also.

---

2 Eric Ferguson, "The Telegraph," 2017, https://www.sermoncentral.com/illustrations/sermon-illustration-eric-ferguson-stories-competition-69570#.

His mind was on other things; he would have to finish his chapter later. As he got up to go to class, he saw Csc standing near the gate. She was staring at him, looking uncomfortable. He walked over to her and said, "Csc, are you OK? You look a little nervous."

Sin watched her bloodshot eyes dart back and forth from him to the gate, though her eyes never actually met his for longer than a split second. "Of course. I'm just going to class," she said with a nervous giggle.

"I'm headed to class myself. I'll walk in with you if you wan—"

"No!"

Sin was startled by such a brash response. Even though he couldn't help thinking she looked and acted overly medicated, he felt some compassion for her, wondering why she usually looked as if she was having a difficult time.

"I can't go in this gate," Csc admitted before she looked down at her shoes.

"Why?"

"I just…can't. I feel undefined here."

"Which gates can you go through?" Sin asked, playing along though he was puzzled. He didn't feel undefined at any of the university gates.

"The gates at 90° and 270°."

"Then why are you hanging around this entrance?" Sin asked carefully.

Csc quickly looked at him before answering. "I like the design on this gate. I just like to look at it. Better go. See ya."

Sin watched her pivot and walk off to head around the circumference of the campus to enter through another gate. *I wonder why she can't go through the gates at 0° and 180°.* He turned his head to look at the design on the gate at 180° she liked so much. He observed the evenly spaced, vertical black iron rails, the single horizontal black iron rail running through the center of the gate, and the black iron U-shaped designs taking up the negative space between the vertical rails, alternating between the U facing upward and then downward.

**CSC X**

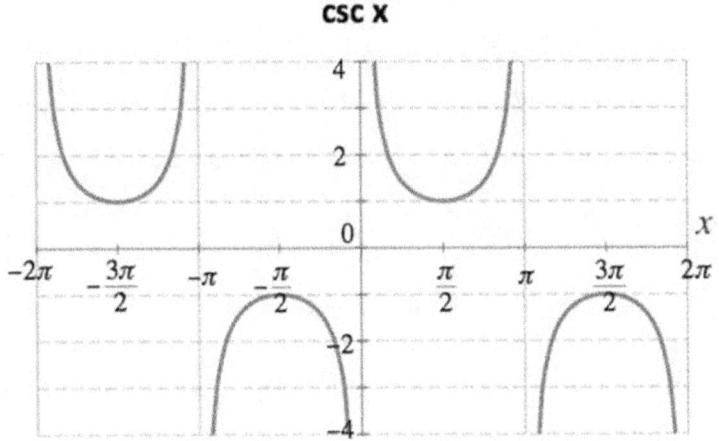

As he began to enter the gate, Sin noticed something he hadn't noticed before. There were tiny engraved markings on the horizontal rail and vertical rails where the two halves of the gate met when the gate was closed. It was as if the markings were indicating a coordinate system, using the horizontal rail and central vertical rails as axes.

After the encounter with Csc, he had gained some insight into Csc's strange behavior. *No wonder Csc likes this gate,* he thought. *This design reflects her personality and how she behaves, but she is asymptotic here and at the 0° gate. Now I understand why she can't enter these two gates. I guess I should take some lessons from "The Telegraph," that story I like so much.*

● ● ●

Csc was walking toward the gate entrance at 270° with her heart pounding. *Sin actually talked to me!* She repeatedly went over the conversation in her mind. She felt like a fool. She had always felt secondary to Sin. She could never be good enough to date him or even to be his friend. *He must think I am a total idiot.* Csc could feel the sting of tears starting to form again but did her best to hold them back.

CHAPTER 7

• • •

TAN WAS SITTING AT A table in the university library located at $7\pi/6$ in the Q3 district, working on her senior project for her degree in architecture. This district was where she felt as if she was in her element, and to her the university's library was a beautiful environment in which to study.

Her senior project was to design a building to be constructed on campus. The university gave the architecture students in their senior year the opportunity to submit a design. One of the designs would be selected for a campus construction project as long as it fulfilled the requirements. She would be graded on aesthetics and curb appeal, accuracy and quality of design, and how effectively she maximized space and minimized cost. This was her favorite way to work, to be given a set of design criteria and to be creative about how she would arrive at the end goal.

Tan's blueprint for this project was coming along nicely. It showed several angles and side measurements for various shapes, which enabled her to determine how much of a specific building material was needed for the space she had. She could then apply current pricing to the material and adjust her budget accordingly. Calculating unknown sides and angles to determine lengths, heights, slopes, surface areas, and volumes—to name a few—was her specialty. She loved working with size and proportion.

Tan used trigonometric definitions, identities, and formulas to create the pieces needed to complete the puzzle as efficiently and as beautifully as possible.

The environment of the library helped her mind flow. She was surrounded by beautiful architecture and design elements with which she identified.

*The Trigonometry Tryst*

At the moment, she was working on calculating the pitch of a portion of roof. Anytime shapes varied from lines that were strictly horizontal or vertical, her favorite skills came into play. She was a pro at calculating slopes and loved to calculate angles of elevation and depression; right triangles were one of her preferred shapes with which to work.

Tan had already calculated the rise and run of the section of roof, so she was ready to calculate the slope and the angle of roof pitch. The rise was five units. The run was twelve units:

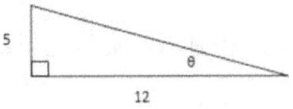

Slope of roof:

$$\tan \theta = \frac{opp}{adj} = \frac{5}{12}$$

Angle of roof pitch:

$$\tan^{-1} \frac{5}{12} = \theta = 22.62°$$

After calculating the surface area of the entire roof and how much roofing material she would need, Tan determined that her material of choice would fit the budget. Wanting to use synthetic-slate roof tiles for durability as well as beauty, she was motivated; everything was falling into place. She just needed the plot survey report for where this building would be built. The address was $5\pi/4$, just next to the library in the Q3 district; the location couldn't have been more convenient for her. She needed to make sure the ground dimensions of the building were appropriate for the plot boundaries and contours, which would be determined from the survey report. She also needed to make sure there was room for the landscape design.

The architecture department and the civil engineering department at the university had an exchange program where graduate students of either department worked on problems or requests as a service to the other department. Since Tan needed a plot survey report, surveying being a specialized function of civil engineering, a graduate student in the civil engineering department would perform the survey and generate the report. The architecture department provided similar services to the civil engineering department, since there were many complementary functions performed between the two departments.

Tan lightly bit the end of her mechanical pencil as a stained-glass window commanded her attention. The library had high ceilings that tapered to dome shapes in some of the rooms. The room she was in had a domed ceiling in the center of the room, with six stained-glass windows spaced evenly around the dome. The perimeter of each window was in the shape of a hexagon. The stained-glass windows were familiar study companions; she had contemplated their geometric beauty thousands of times.

Within the boundary of the hexagonal perimeter of one of the stained-glass windows, a geometric world came alive where shapes composed of shapes could be found in a never-ending loop of discovery, if one were willing to look deeply.

Tan liked first to pick out the six equilateral triangles making up the hexagon. Each of the equilateral triangles was composed of smaller equilateral triangles positioned to create smaller hexagons. With the enhancement of several deliberate colors designed into the piece, the color contrast brought out shapes that might have gone undetected to the casual observer.

When the sunlight filtered through the window, as it did now, the six-pointed star revealed itself with its lighter color distinction. Tan never tired of the beauty she found in this geometry.

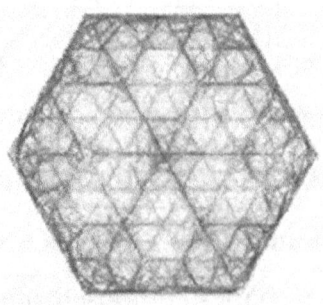

The six-pointed star itself could be created with the overlay of two equilateral triangles positioned to form a hexagon at the center of the star. Tan's fascination with this particular window was that every wonder emerging in this design was fundamentally made up of equilateral triangles. One of Tan's favorite and most used triangles, the 30-60-90 special right triangle, was also derived from the equilateral triangle:

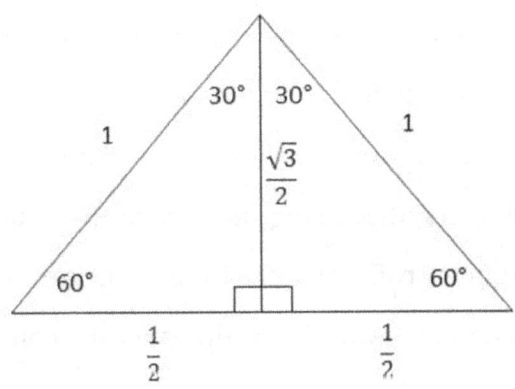

She was about to focus on her project again and use the 30-60-90 special right triangle to make another calculation, when she noticed another way she could use the triangle. She most liked the ratio of $\tan\theta = $ *opp/adj*, which was her right-triangle definition. However, she noticed that since she already had the measurement of the hypotenuse of the right triangle she was working with, another ratio would work better.

She could use the ratio *opp/hyp* to find the side measurement she wanted. This ratio was familiar to her. She used it regularly when solving right triangles, but there was something else. She remembered seeing this right-triangle definition on the mystery profile Cos was interested in.

Tan's mind drifted toward Cos's having a profile on Unit Match. She was aware of Unit Match and heard other students discussing it constantly—whom they met, what new profiles had been posted, and whom they hoped to date. However, Tan always let the conversation go in one ear and out the other. She would smile politely when people talked with her about what they last saw on Unit Match, but Tan never had any interest in getting on the website. She was more interested in her studies.

She had goals she wanted to accomplish and didn't want to waste time on social media or online matchmaking sites, because she didn't sense that any of the conversations were worth her time.

She smiled at the thought of Cos's anticipation of finding someone to date on Unit Match. She knew Cos didn't like to be alone, but she admired her discernment versus desperation when choosing those she wanted to be with. Cos was attracted to edifying people. *I am attracted to edifying people, but I also love to be alone. I love the time to study, read, think, and plan.*

Although Tan enjoyed her own company, this wasn't because of arrogance or shyness. She wanted her time to be productive and purposeful. Simply hanging out for long periods was excruciating and left her feeling bored and impatient.

Tan knew this about herself, and it concerned her at times. While Cos always told her she was as solid as a rock, Tan had some anxiety about her desire to be alone and her principled and uncompromising nature. She did need some balance. She needed to be around people for short periods and then was happy to return to her solitary activities.

Tan loved to be with people one-on-one. She enjoyed having friends. Going out to lunch or dinner and talking about real things was enjoyable to her. Then it was time to move on to the next thing.

*If I like to be alone so much, do I have it in me to be in a family? To work together well with and love a spouse? These are important events in a human life. Will purposely avoiding these things altogether cause me pain later? I should ask Cos if it is an effort to want to be around people all the time—if it is a natural desire in her, or if it takes work. I do know that if I ended up with someone, the match would need to be a strong one, deep and fulfilling, someone I really enjoyed talking with and partnering with. A match with someone I want to be with more than I want to be alone.*

Thinking about Cos's attitude sparked an interest in Tan, and she was curious for the first time to see whether she would be attracted to any of the profiles on Unit Match. She hadn't once looked at the site.

Entering the password to get on the library's Wi-Fi connection from her laptop, she allowed herself a rare distraction from her scheduled study time to check out Unit Match. The home page was full of ads and testimonials from students who had met boyfriends or girlfriends through the site and were happily dating. Of course, the perfect marketing lines—"Your perfect match is waiting for you!" and "Meet your complement here!"—shouted at her, as the advertising slogans were designed with large, bold, and animated fonts to make sure they were the first things she saw and digested. *How could anyone resist going further into the site?* Tan chuckled to herself as she subtly rolled her eyes.

Tan realized she would have to create an account if she wanted to view anyone's profile, although it wasn't required that she fill out a profile. Her curiosity got the better of her, and she decided to create an account on Unit Match and fill out a profile anyway. The questions were easier to answer than she thought they would be. Pretty black and white:

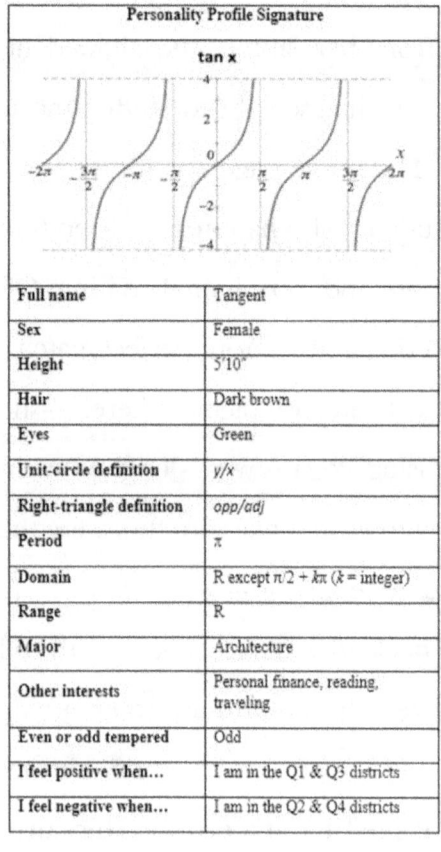

| Personality Profile Signature | |
|---|---|
| tan x | |
| Full name | Tangent |
| Sex | Female |
| Height | 5'10" |
| Hair | Dark brown |
| Eyes | Green |
| Unit-circle definition | $y/x$ |
| Right-triangle definition | $opp/adj$ |
| Period | $\pi$ |
| Domain | R except $\pi/2 + k\pi$ ($k$ = integer) |
| Range | R |
| Major | Architecture |
| Other interests | Personal finance, reading, traveling |
| Even or odd tempered | Odd |
| I feel positive when... | I am in the Q1 & Q3 districts |
| I feel negative when... | I am in the Q2 & Q4 districts |

Tan decided to look at Cos's profile first before viewing others, to see how she described herself:

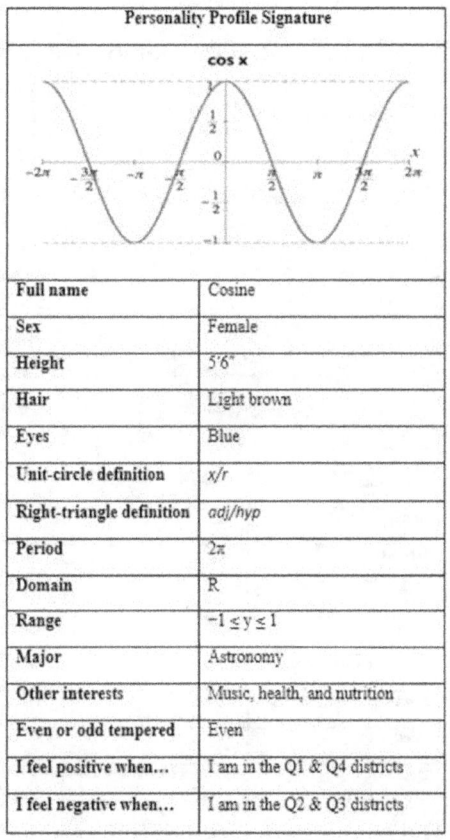

*Interesting. We get along so well, but looking at our profiles, we have nothing in common. I wonder if there is someone out there who has a profile similar to mine.*

As Tan searched for a match to her own profile characteristics, she found one that had the same range she did of all real numbers, R. This person was also odd tempered, as she was, and liked the same districts she did:

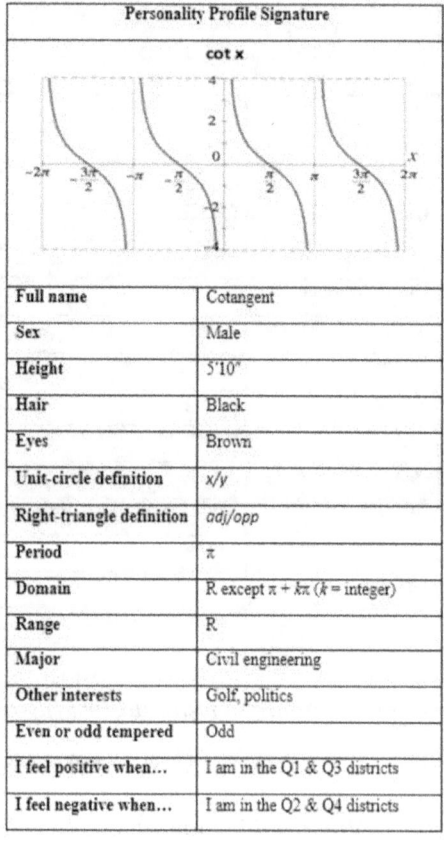

| Personality Profile Signature | |
|---|---|
| cot x | |
| Full name | Cotangent |
| Sex | Male |
| Height | 5'10" |
| Hair | Black |
| Eyes | Brown |
| Unit-circle definition | x/y |
| Right-triangle definition | adj/opp |
| Period | $\pi$ |
| Domain | R except $\pi + k\pi$ ($k$ = integer) |
| Range | R |
| Major | Civil engineering |
| Other interests | Golf, politics |
| Even or odd tempered | Odd |
| I feel positive when… | I am in the Q1 & Q3 districts |
| I feel negative when… | I am in the Q2 & Q4 districts |

*The Trigonometry Tryst*

Tan looked up from her computer screen to stretch her back, when she noticed the guy at the next table looking at her. He didn't look away when they made eye contact, so she glanced back at her computer screen for a few seconds. When she looked up again, he was still looking at her. This time he smiled, got up out of his chair, and walked over to her.

"Sorry, I didn't mean to make you feel uncomfortable. I'm Cot," he said as he held out his hand to shake hers.

"I'm Tan."

"Nice to meet you. You look pretty absorbed in what you're working on. I don't mean to bother you. I don't study in this room all the time, but you looked familiar. That's why I was looking at you."

Tan noticed he was glancing at her computer screen. She felt embarrassed that he saw she was spending time on Unit Match. She always perceived Unit Match as a silly pursuit and assumed others would view her the same if they knew she spent time on the website.

"Unit Match is an easy diversion from the important stuff, isn't it?" Cot said with an amused look. "Any interesting profiles?"

"Well, you might not believe this, but this is the first time I've ever spent time on Unit Match, so I'm not sure what to tell you." Tan watched Cot's steady, deep-set chocolate-brown eyes hold his expression as she asked, "How about you? Have you found anything worthwhile on this website?" She felt that uneasy feeling of her privacy being invaded—or maybe her pride was taking the lead. At the same time, she noticed Cot's calm, soft voice.

"Oh, I've surfed through the site a time or two, but nothing has stood out so far," he said. "Well, I'd better let you get back to your studies. Architecture major?" Cot asked, catching a glimpse of Tan's blueprint.

"Yes, senior project. It was nice to meet you." Tan liked that his eyes looked intelligent.

"Same here. I'm sure I'll see you here again." Cot turned to walk back to his table. He smiled to himself, since it was his profile Tan was looking at on Unit Match. *She must not have looked at it long enough to have made the connection between the name on my profile and the guy standing right in front of her.*

CHAPTER 8

● ● ●

Cos began walking back to her apartment after her class was over, thinking how glad she was that Tan was her new roommate. The Q1 district was the best place to live on campus, and with a roommate whom she got along well with to share the rent, she felt her circumstances were ideal.

Cos was walking counterclockwise from the Q4 district, where most of her astronomy classes were, to the Q1 district, feeling positive about things. She had a bounce in her step as she looked up at the sky, where her eyes landed on a large, beautiful full moon. Even though it was a sunny day at noon, the smooth blue sky served as a striking backdrop to contrast with the magnificent white disk.

Cos had expected to see a full moon on this day, February 10, as she was always aware of the fraction of the moon illuminated on any day. She regularly looked up the schedule produced by the Naval Observatory that showed when the moon would be waning or waxing on any given day of the month.

Having just come from her Astronomical Observation and Data Analysis class lecture, Cos realized she had never graphed the data available from the Naval Observatory to see how the moon's illumination behaved. *It's periodic for sure, since the moon's waxing-and-waning cycle repeats itself every month.*

Cos decided she would look up the available data from the Naval Observatory for this year and graph it.[3]

---

[3] "Fraction of the Moon Illuminated," US Naval Observatory, February 15, 2017, http://aa.usno.navy.mil/data/docs/MoonFraction.php.

She stopped short, thinking she would go over to the library in the Q3 district to do this work instead of going back to her apartment. *Tan should be there.*

Cos walked into the university library, but before entering the common study areas, she liked to spend a moment in the high-ceilinged foyer viewing the art panels lining the walls. Even though the Q3 district was not her favorite, one astronomical panel captured her attention. She was drawn to artist renditions of the subject she loved most. Of course, the library panel depicting the phases of the moon held particular interest for her.

She walked up to the panel for a moment to mentally greet it. She smiled, noticing that the moon was waxing clockwise. *You can always tell whether the moon is waxing or waning. Illumination grows or waxes from the right side.*

Pleased to find Tan sitting at one of the tables, Cos walked over to her, knowing Tan loved this library in the Q3 district and spent a lot of time here. Tan was surprised to see Cos, who took the seat next to her.

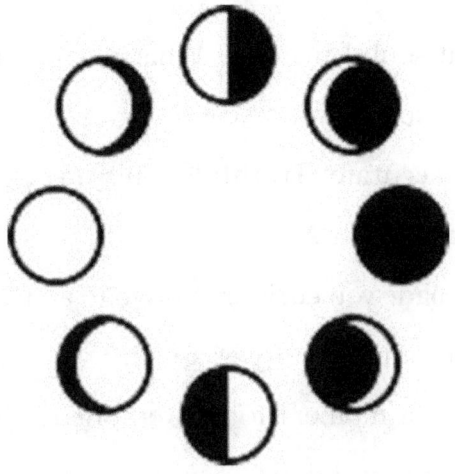

"I haven't seen you here in months!" Tan said with a wide smile.

"I know. You know me well enough to know that the Q3 district is not my favorite, but this is a beautiful library, and I came because I knew you would be here."

At that moment, Cos glanced at Tan's computer screen, and her eyes widened.

"What?" Cos said. "The sagacious Tan on the Unit Match website that wastes so much of students' time?"

"OK, I deserve that. My curiosity got the better of me. I even looked up your profile."

"Does it sound like me? What do you think?" Cos asked with excitement.

"It was accurate. Truthful. You won't get any complaints." Tan winked.

"What made you curious enough to actually take the time to look at the site?" Cos asked.

Tan took a moment to answer. She sat back in her seat. "You, I guess. I don't know. You're not afraid to put yourself out there in this way, while I feel a little more reserved about sharing personal things with just anyone. Plus, sometimes it's nice to relate to what everyone is talking about, isn't it?" Tan asked thoughtfully.

Cos watched her friend share her thoughts. Tan didn't usually share her vulnerable side, but Cos wasn't naïve enough to assume she didn't have one.

Tan was solid and principled, and she didn't waste time. Those who criticized her perceived her to be uncompromising and stubborn. That was OK with Tan. She knew where she was headed and viewed her critics

*The Trigonometry Tryst*

as wanting to hold her back. Cos considered it a privilege anytime Tan opened up to her.

"Do you think you'll put your profile on here?" Cos asked. "Hey, I'll help you do it now."

Tan slowly smiled and pulled up the profile she had already completed. She turned her computer so Cos could look at it. "I even found a profile that had some similarities to mine, but not as many as the nameless match you found when you were helping me move. Did you ever find out who it was?"

"No, I haven't looked too hard yet," Cos said.

"So what do you think about my profile?" Tan asked.

"I'm still shocked but excited that you created an account," Cos said, looking through Tan's profile. "I don't see anything that doesn't describe you."

Cot smiled to himself again at the next table over as he strained to overhear Cos and Tan's conversation. It was his profile Tan liked. He pulled up Unit Match and searched for Tan's profile, overhearing she had just created one. There were some complementary characteristics and a couple of similarities, such as their ranges and the districts in which they felt positive or negative. They were both odd tempered; however, he could see they were reciprocals of each other in just about everything else. *I like her, though. There's something unique about her.*

● ● ●

When Cos and Tan began to focus on their studies in the library, Cos pulled up the Internet to view the Naval Observatory's 2017 data to determine the fraction of the moon illuminated daily for each month. She decided to transfer the first six months of the year of calculated data to a spreadsheet and chart it. She could have predicted what she saw, and it pleased her: a sinusoid or sine wave. She knew that the phases of the moon and its path around the earth would be periodic, smooth, and repetitive, but she hadn't proved to herself which wave form this periodic phenomenon would take. She was glad it was a sine wave:

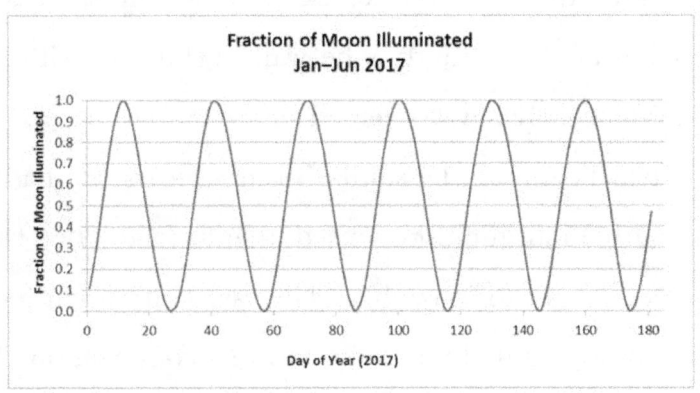

A time-dependent sine wave function takes the form of

$$S(t) = A\sin(\omega t + \varphi) + \beta \text{ for } \omega > 0.$$

- $A$ = amplitude (displacement from baseline to max and min values)
- $\omega$ = angular frequency = $2\pi f$ ($f$ = ordinary frequency – $\frac{1}{T}$)
- $T$ = period (the length of time it takes to complete one cycle of the sinusoid)
- $\varphi$ = phase (horizontal phase shift = $-\frac{\phi}{\omega}$)
- $\beta$ = vertical shift or baseline

Looking at the wave form, she suddenly started to think of Sin. *I wonder if he likes this kind of thing also.* She pictured Sin with his backpack slung over his shoulder, looking at the ground as he walked, but not without confidence. He always seemed so mysterious to her. Yet she felt they would be similar in many ways. He had a strength about him, she sensed, that attracted her; she wanted to get to know him. He wasn't the usual jock who always wanted a date but no real conversation. As even tempered as she was, she was bored with that sort, wanting someone with internal substance.

As she daydreamed, Cos's thoughts went from Sin to the puzzling post on Unit Match Tan had just reminded her about. *How strange that this person wants to meet at a particular time and date but doesn't give a location. Or even his name, for that matter! Tan was right; why waste my time on such an obtuse Invitation?*

However, reasoning was not taking her mind off the Invitation. There were so many similarities in this person's profile to hers. Maybe he wasn't finished filling out the Invitation to Meet. *Maybe I should send him a message to let him know he did not give a location to meet and no one even knows his name.*

As these thoughts went through her head, Cos drifted back to looking at the wave form in front of her. Suddenly, it struck her that if she added a phase shift, the wave equation would become

$$C(t) = A\cos(\omega t + \varphi) + \beta \text{ for } \omega > 0,$$

which she found even more satisfying. She remembered that cosine and sine functions could be used to model many natural behaviors and that the cosine function is also a sinusoid, or sine wave, with a phase shift.

Cos smiled and then allowed herself to be diverted again as she decided to pull up the ad with the missing information:

| Personality Profile Signature ||
|---|---|
| | *(graph of sine function from $-2\pi$ to $2\pi$, amplitude 1)* |
| Full name | |
| Sex | Male |
| Height | 6'2" |
| Hair | Dark brown |
| Eyes | Brown |
| Unit-circle definition | *y/r* |
| Right-triangle definition | *opp/hyp* |
| Period | $2\pi$ |
| Domain | R |
| Range | $-1 \le y \le 1$ |
| Major | Electrical engineering |
| Other interests | Leadership, teaching |
| Even or odd tempered | Odd |
| I feel positive when… | I am in the Q1 & Q2 districts |
| I feel negative when… | I am in the Q3 & Q4 districts |

| Invitation to Meet | |
|---|---|
| Date | April 30, 2017 |
| Time | 8:07 p.m. |
| Place | |

"I think I'll let this guy know he didn't tell anyone where to meet," Cos said.

Looking away from her blueprint, Tan saw that Cos was looking at the mystery profile again. "Go for it," Tan encouraged.

"You could contact the person with the profile you kind of liked," Cos said.

"Don't push me too far in one day!" Tan joked. "At least I looked at the website and even created a profile today."

Cos was so drawn toward this profile that she decided she would send the owner a message. There was a field under his profile to send a message to his e-mail. Unfortunately, his e-mail address gave no indication of who he was either. Cos sent a message:

Hi!

As I was browsing through Unit Match, I ran across your profile and saw this Invitation to Meet. I was just making sure you knew that your name and location for the meeting were missing. I thought you would want to know. Have a good day!

Cos

● ● ●

As Cos was finally crawling into bed, she checked her phone for any text messages and e-mails. She had a response to her message:

Cos,

I'm glad you're interested. I challenge you to a fun experiment. Figure out the location…if you can. I look forward to meeting you!

*The Trigonometry Tryst*

Cos turned her lamp off and lay in the dark, staring up at the ceiling, a little astounded. "OK?" she said aloud. She didn't realize she had overlooked a hyperlink that was just underneath his message.

**CHAPTER 9**

• • •

Csc contemplated the design in the 180° gate again, as she regularly did. *Sin probably won't be on the bus today. The weather is good.*

Csc was becoming aware of some things about herself she wasn't used to. Her feelings always felt so automatic, as if she didn't have any control over them. Emotions would come and go. Highs and lows recurred several times in one day. Just like the design in the gate.

Wanting to touch the gate and run her fingers across the up-and-down U-shaped designs, she knew she couldn't. She could only come infinitely close but never touch. She was asymptotic here and at the 0° gate.

Csc thought she would be happy doing favors for Sec, making him happy. She was the right-hand person for the star basketball player at the university. Even then, the emotional highs she felt when Sec asked her to do something for him, the winks and pats on the back he rewarded her with, could merge into a low, and she'd end up crying herself to sleep at night, not knowing why. She was unhappy, but she didn't know what to do about it. She didn't know what it was like to be loved. Not having been loved as a child, always having been told she was secondary, was emotionally dangerous. It meant overreacting to accept even a hint of love or acceptance from anyone and being unable to discern, without painful experiences and more maturity, whether that person had her best interest in mind.

Csc was aware of her people-pleasing weakness and that she came off as being unselfish, always doing what other people wanted her to do—but she knew she was, ironically, selfish. Most people pleasers, she knew, had emotional issues, doing what other people wanted them to do because they didn't want someone to be mad at them; they were always looking out for their own insecure feelings. They also thought they were somehow not allowed to state their opinions. They had opinions but didn't share them. Csc knew this intellectually. She could see it, but psychologically, she couldn't act on it.

A truly emotionally healthy person who could stand up for herself in a mature way while being aware of and truly wanting to help other people was a rarity. Someone operating on this emotional level was not a people pleaser. She was a leader.

Csc brushed a blond curl away from her hazel eyes, which were following the design in the gate. Upward-facing U, downward-facing U, upward-facing U, downward-facing U. *Ups and downs. Highs and lows.*

CHAPTER 10

● ● ●

Tan wanted to finish her senior project, but she was annoyed the survey report wasn't finished yet. She needed to confirm that the design of her building foundation would be appropriate for the space she had within the boundaries of the plot, as well as make sure there was room for the landscape design. She had seen the plot reserved for the building next to the library and was confident, but she liked to do things by the book. Being organized and doing things right the first time, she experienced less stress and worry. Besides, the survey report needed to be submitted with her design proposal.

On her way to the library, Tan decided to stop by the engineering building to check on the timing of the survey report. As the elevator doors opened on the fourth floor of the building, where the civil engineering department was located, Tan stepped off the elevator and saw the black-haired guy walking down the corridor away from her. *What was his name?* She tried to remember the guy who had introduced himself to her in the library. *Cot! That's it. I don't think we talked about what he was majoring in.* She went to the department office.

"Cot hasn't started on the survey yet," a grad student informed her after checking their work log.

Tan expended some effort to keep her composure. "This survey request has been on your department's log for four months now," she said carefully.

The grad student wanted to run away as he experienced Tan's piercing green eyes boring into him, even though he knew he was only the messenger.

"Did you say someone named Cot had been assigned to do this survey? Does he have black hair?" Tan asked stonily.

"Yes…"

"Look. I know this is not your fault personally, but I really need this report to finish my senior project. How can I find out when it will be completed?" Tan asked with authority.

"I can tell Cot you came by…" the grad student said.

"That's not good enough. Which professor does he work for?"

As Tan received the information she needed and left the office, the grad student and the department secretary, who had watched the interaction, looked at each other, not understanding how someone developed the kind of backbone they had just witnessed in Tan.

Tan was hoping Cot's professor wasn't teaching a class at the moment. She was about to knock on her office door.

"Are you lost? Shouldn't you be over in the architecture building?"

Tan turned around to see Cot with a smile on his face. "Hi, Cot. Nice to see you again. Actually, I should be in this building, because I've been waiting for a survey report for four months now from the civil engineering department for my senior project, and I've just been informed it has not even been started yet."

"I guess they had all better get into gear, hadn't they?" Cot seemed oblivious to the problem.

Tan looked at Cot with her usual controlled manner, but without a hint of a smile or any kind of indication Cot was going to be let off the hook. "So you're a civil engineering major?"

"Yep. Grad student."

"I understand you are the grad student assigned to the survey I've been waiting for," Tan said, watching him with curiosity to see which way he was going to go with this.

Cot tried a false laugh to smooth over the situation. Noticing no reaction from Tan, except for the steady stare in her green eyes that grew more intense, he said with some annoyance, "You don't have to be so critical. I'll take care of it and let you know as soon as it's done. Can I get an e-mail address or phone number to contact you? Give me until the end of the week."

Tan watched Cot as he walked away with her e-mail address and heard him breathe out an impatient sigh. She shook her head. The irony never ceased to amaze her when she observed irresponsible people blaming others for their own failures. *Somehow, they think this works! They think they are being subtle when the game they are playing is as obvious to others as the nose on their face. As if I had no right to question the timing of the completion of the survey report. They try to make a fool out of you, but it's clear who the fool is! Integrity is not a personality trait; it's a character trait. This is the guy whose eyes I thought were intelligent and whose voice was comforting?*

If Tan were going to lose her calm demeanor, this kind of behavior from others pushed her buttons the most. Incredibly unimpressed with Cot, Tan headed over to the library.

● ● ●

Cot hauled his surveying tripod around the campus unit circle to the Q3 district, where the starting boundary point of the plot of land for Tan's project began.

To add to the agitation he was feeling, the wind was blowing hard. Wind could always be a natural source of error in a survey measurement. The department had students use outdated equipment as practice to manually calculate their work. This helped students better comprehend the concepts involved in the procedure. Updated digital equipment could be used further along in the curriculum. Cot would have preferred to do the survey on another day without strong wind, but he felt completing the survey quickly outweighed the potential measurement error. He began taking the measurements to determine the area of the plot of land designated for the campus building.

Avoiding conflict as much as possible, Cot was observant of other people and wanted to be respected and looked up to as a leader. He was used to charming anyone out of just about anything. He was usually calm and pleasant, but complacent. *Tan is too principled and needs to lighten up.* He couldn't forget those green eyes on that face of steel that had penetrated him as though she was able to uncover his real intent. He was used to keeping commitments only for perception's sake, not because it was the right thing to do.

The Q3 district was Cot's favorite district in addition to Q1; everyone liked Q1, though. However, he wasn't feeling on top of his game today. The image of Tan's piercing green eyes drilling through him continued to stir up emotions of hurt and anxiety he couldn't shake. The image flashed in his mind repeatedly, telling him he was a failure. His usual act wasn't working.

Cot's top strengths had let him down this time. He depended on his calm demeanor and diplomatic tendencies to prevent confrontation and draw people to him.

"Jeez," Cot muttered as he staked out the corners of the plot of land with the information he was given: the four boundary points, determined from viewing the surveys from the neighboring plots with existing buildings.

Cot could see that the plot of land was not going to be a perfect square or even a rectangle, which would have made his job easier; working with ninety-degree corners was a breeze. Since the plot of land had a quadrilateral shape with four unequal angle measurements, he was going to need to use some complex calculations to solve the quadrilateral; he needed to find the needed angle and side measurements in order to come up with the area of the plot of land.

Shooting the distances between one boundary point, A, and its two adjacent boundary points, B and D, and measuring the angle A between these two distances, Cot sat down and drew a diagram. His focus was off as his thoughts bounced back and forth between finding the area of the plot of land and Tan.

*I like Tan. Why did she have to talk to me like that?*

Cot decided to use non-right-triangle trigonometry, as he was not working with any right triangles. With the information he was given, he chose to cut the quadrilateral into two triangles so he could sum the areas of each of the oblique triangles to get the total area of the plot:

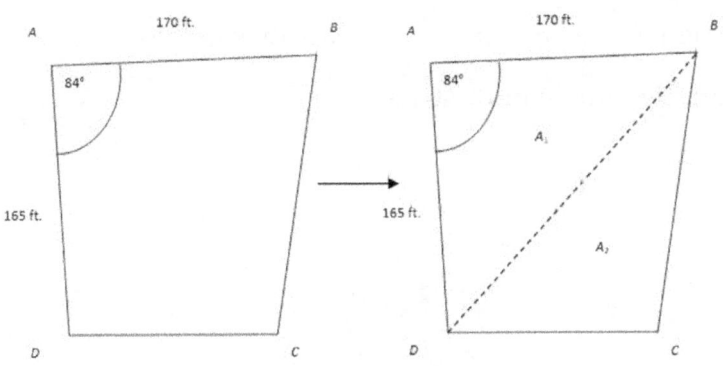

*I'll win her over. I always win people over.*

Knowing the angle measurement at boundary point A and the measurements of the two adjacent sides now, Cot knew he could use the law of cosines to find the third side of the triangle:

$$a^2 = b^2 + d^2 - 2bd\cos A$$
$$a^2 = (165)^2 + (170)^2 - 2(165)(170)\cos(84°)$$
$$a = 224.2 \text{ ft.}$$

*She needs to relax.*

Now that Cot had enough information in the first triangle, he could use the law of sines to determine another angle in the triangle:

$$\frac{\sin A}{a} = \frac{\sin D}{d}$$

$$\frac{\sin 84°}{224.2} = \frac{\sin D}{170}$$

$$D = 49°$$

The third angle in this triangle, B, was easy to determine. Cot always remembered the rhyme:

> All angles that a triangle sees
> Must add up to 180°.[4]

Cot then measured angle C and shot the distances between this angle and boundary points B and D.

Although Cot could have used the law of sines to find the remaining angles in the second triangle, he already

---

[4] J. A. Bailey, *MathOdes: Etching Math in Memory: Geometry* (Festus, MO: MathOdes Company, 2013), 8.

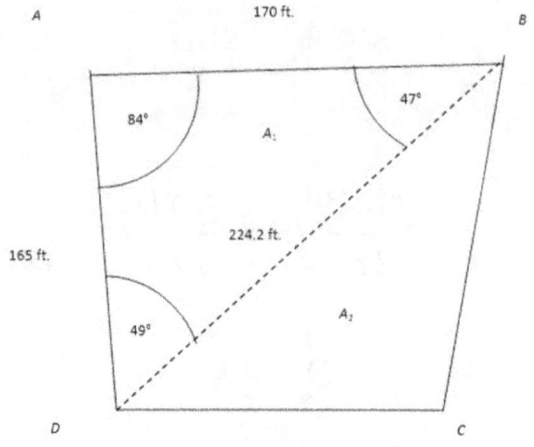

had enough information to find the area of each of the triangles by using this formula:

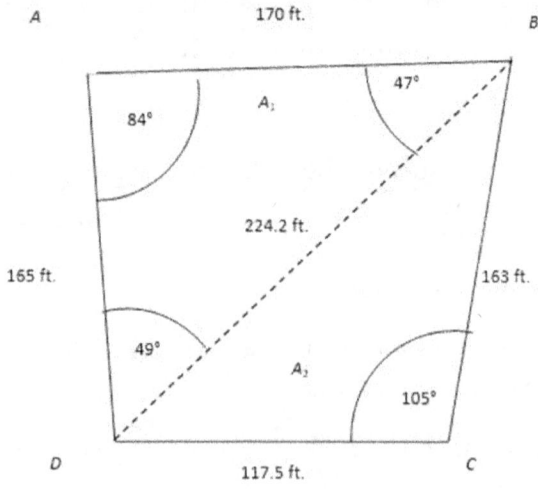

$$A_1 = \frac{1}{2}bd\sin A$$

$$A_1 = \frac{1}{2}(165)(170)\sin(84°)$$

$$A_1 = 13{,}948 \text{ ft.}^2$$

Using the same formula to find the area of the second triangle, Cot calculated the following:

$$A_2 = \frac{1}{2}(117.5)(163)\sin(105°)$$

$$A_2 = 9{,}250 \text{ ft.}^2$$

The total area of the plot of land then became this:

$$\text{Total area} = A_1 + A_2 = 23{,}198 \text{ ft.}^2$$

Tan now had the area of her plot of land for her building design:

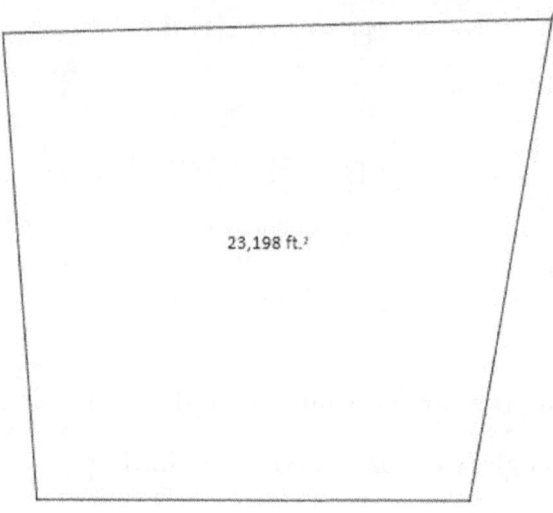

*She'll burn bridges if that is her style of relating to people when things don't go her way.*

At least the slope of the plot he measured was in the range considered flat, which made more complex surveying techniques unnecessary.

*Be patient. I can influence her.*

When Cot was measuring the slope of the plot, his impaired focus had made him careless, and the wind caused his tripod to become unstable. His previous training had taught him that a surveyor is responsible for making needed adjustments when natural phenomena occur. If he had prepared properly and had a clear mind, he would have stabilized the tripod with sandbags or another effective option he needed to consider on a windy day such as this one. But he wrote down a measurement without paying attention to the level indicator, which would have told him his tripod was not level.

● ● ●

Tan's annoyance with Cot stuck with her. *Unbelievable. It never ceases to amaze me that when people break a commitment or drop the ball, they look at you like* you *have the problem. Their credibility could be saved if they would just own up to it.*

One of Tan's favorite leadership authors wrote,

In the first six months of a relationship, whether it's personal or professional, one to one, or leader to follower…the first six months, communication overrides credibility. After six months, credibility overrides communication.[5]

*Cot clearly thinks he is entitled to have credibility regardless of how he behaves. Unbelievable!*

---

5 John C. Maxwell, *Everyone Communicates Few Connect: What the Most Effective People Do Differently* (Thomas Nelson, 2010), audiobook.

CHAPTER 11

● ● ●

Sec finished basketball practice in Q4 and headed counterclockwise with some of his teammates to the 0° gate. Seeing Cos up ahead at the gate, he watched as she confidently talked with her friends with such ease. He never saw her with any of the girls he usually went out with. *Why is that?* She was just as good looking as the other girls, if not more so. That's why he wanted a date. Why did she seem so distant when other girls he knew would die to go out with him?

Cos seemed relaxed and gracious, not trying to impress anyone. Sec watched as she listened with genuine interest and conversed with patience and friendliness. Her light-brown short hairstyle was professional, and her stylish but modest way of dressing upgraded her further in his eyes. Why was he so attracted to her? He subconsciously knew he wasn't good enough for her; he was secondary to her.

Cos could feel his eyes on her even though Sec's teammates were making a ruckus at the gate. Sec was leaning against the black iron at the 0° gate. He knew she knew he was a star athlete. He had initiated a conversation with her once. Sometimes she passed him at the 0° and 180° gates, but she didn't acknowledge him. He was attractive physically, but her gut told her to be wary of this one. She sensed some shiftiness about him. Was it the lack of depth in his eyes? Was it his unnatural stance, which seemed all for show?

Sec watched Cos suddenly become uneasy. Quite a change from the effortless grace and conversation just a few minutes ago. Just then, Cos looked directly at him. Their eyes met. He froze and couldn't bring himself to smile at her. There was only a blank expression on her face as well. She left her friends and walked through the gate. He still felt his jaw clamped together as a reaction to overhearing Cos mention an Invitation to Meet on Unit Match she was interested in. He didn't overhear the name of the person whom she was interested in, but he had gathered enough information he could use.

● ● ●

Trying to shake off the agitation she felt as she walked around the circumference of the campus, Cos turned her thoughts toward the mysterious Invitation. Two different feelings. The Invitation to Meet made her smile. Sec made her skin crawl. She really wanted to meet this mystery person. How could she find out where to meet him? Evidently, the mystery person thought she could figure it out.

CHAPTER 12

● ● ●

Csc sat on a stone bench in some shade facing one side of the clock tower at the center of the university. Her cybersecurity class had just let out. She was meeting Sec at the café again later, but she wanted to sit and think for a minute.

Sec had just sent her a text telling her to find out who had posted an Invitation to Meet with no name or location. He had found out Cos was interested in this person's profile, and he was going to interfere.

It was sunny, which Csc was grateful for; she could wear sunglasses and not feel that her emotions were exposed to everyone walking by. She knew her eyes would be bloodshot and on the verge of tears. When insecurity dominates thoughts, the only one in the world being watched, criticized, pointed at, and shunned is the one dwelling on them. It's the inversion of being a diamond in the dirt. It's like being a clod of dirt in a sea of what appears to be perfectly well-accepted and beautiful human diamonds. What was it like to blend into that sea of beautiful, shimmering diamonds? To be a part of something? What was it like to be accepted? What was it like not to feel like a freak?

A part of her cybersecurity class today had sparked an idea to accomplish the favor Sec asked of her this time. Ethics and morals mean nothing when one is so starved for human connection and acceptance. The light of her conscience regarding this matter was dim, overshadowed by her insatiable appetite for approval from others.

A blond curl fell across the right lens of her sunglasses as she looked across to where the clock tower stood and watched students pass by in various radial directions from the clock; some used the engraved cement compass to determine which radial direction to walk in.

She saw Cos approaching, walking with an unusual seriousness about her, looking at the clock tower. *Does this have to do with Sec? Should I ask her about the Invitation to Meet she is interested in? No. Too obvious.* Csc was going to behave in her habitual way: Pretend not to notice. Take no initiative to say hi to Cos. Withdraw and close in on herself. She looked down, pretending to be busy, again grateful for the shield of the sunglasses. She would look up again when she was sure Cos had passed by.

"Hi, Csc!"

Csc looked up to see Cos smiling down at her. "Oh, hi, Cos. How are you?"

"Good. I saw you and just wanted to say hi. Hope you're doing well."

"Never better," Csc replied, not realizing she was unconsciously wringing her hands.

"Well, good to see you. Come see Tan and me sometime. You're always welcome."

"Thanks." Csc watched Cos walk off, grateful for the conversation but always embarrassed, thinking she hadn't said something right. After all, Tan had left her as a roommate and moved in with Cos. No amount of reasoning could take away the feeling of personal rejection.

Csc watched people walking with purpose all around her, thinking they were problem-free and accepted by others with hugs, invitations, approving comments, and rewards. In her distorted perception, she was the only one who passed through this life in misery.

She pulled her laptop out. With no shortage of WAPs covering the university's Wi-Fi network, she was able to use her knowledge and skill to find some answers on Unit Match. Who had sent the Invitation to Meet without a name, the Invitation to Meet Sec had overheard Cos expressing an interest in at the 0° gate? When she saw who it was, she knew Sec would blow a gasket. His jealousy would rage. But *she* would not be rejected. Her actions would make him accept her even more.

Before noticing Csc sitting on the bench, Cos watched the clock tower as she got closer to it. She watched the second hand make its way around the clock. She heard the university bus drive by at a rate in sync with the second hand of the clock. For some reason, more than just the time registered in her awareness. The clock's second hand made one revolution every sixty seconds. In her field of vision, she could see the university bus making its way around the circumference of the university. *Interesting.* She realized that the bus was traveling at the same rate around the university circle as the second hand was moving around the clock face.

*The second hand and the bus must both have the same angular velocity. What a coincidence. Of course, the bus needs to stop at each gate, while the second hand moves continuously.*

CHAPTER 13

● ● ●

Sec was walking by the sports stadium in the Q4 district, thinking about how he was going to use Csc to further the next steps in his agenda. How he loved to spot the weaknesses in others and use them for his own gain. He felt triumphant as he walked past the university's basketball court, thinking of his conquest over Csc. She had fallen right into his hands. This was Sec's version of success, manipulating those around him to serve him with no thought for their welfare.

He walked with his head held high as he entered the locker room to change for basketball practice. How could he fail at anything now? He was the most popular shooting guard in the university's history—even more popular than the point guard. Sec's skill at shooting baskets and his average height—he was not short but was taller than the point guard—made him perfect for this position.

Sec spent a great deal of time analyzing perimeter shots and at which angles to shoot depending on his position on the court. When Sec shot free throws, he knew that if the ball's pathway took the form of an arc in the air, he had the best possible chance of making a basket because of the margin of error available for the ball to go into the basket.

According to physicist Peter Brancazio, if a ball is shot at a forty-five-degree angle, the ball will arc in the air in a way that delivers it right through the basket. This angle can change slightly depending on the height of the player or the distance from the basket. If the ball is shot at an angle of thirty-two degrees or less, the odds are not very good for making a basket, because the margin of error between the size of the ball and the basket decreases. The basket becomes more elliptical, lowering the chances of the shooter making a basket.

Sec's free throw record was 90 percent, making him a valuable player.

As he was majoring in marketing, but with a minor in physics, Sec was able to combine the physics of shooting a basketball with his love of sports and draw the praise of crowds with his superiority. This was how successful people made it to the top and survived. But his dark posture of competition and manipulation, along with his lack of awareness of the principles of lasting success, resulted in only temporary success and the need for constant scheming to keep himself on top at the expense of others.

Feeling invincible, he decided to make a move for Cos. After all, he had everyone else in the palm of his hand; he could have any girl he wanted, right? Why wouldn't Cos want to go out with him? He was the star basketball player. Good looking. Smart. And he had business ambition, which meant money.

Cos was hot. He needed a trophy girlfriend to impress the crowds and make his friends jealous. There was no better feeling of victory than being envied by others.

All these thoughts swirled around in his mind as he showered in the locker room after practice.

"Hey, man, I've got a hot date tonight," one of Sec's teammates shouted over the running water in the adjacent shower stall. "Met her on Unit Match. You should check it out. Even though you probably have enough girls lined up to go out with you to last you until you graduate. If I could be so lucky."

This was just what Sec loved to hear. He didn't tell his teammate, but he had been studying Cos's profile on Unit Match for some time now.

His past experiences told him he was a desired date and had no need of Unit Match, but for some reason he didn't understand, a sense of inferiority bubbled up inside him when he thought about Cos. He felt secondary to her. This was not a feeling he was used to. He tried to ignore the way she had looked at him at the 0° gate. He could make himself feel better if he just assumed she was shy.

Even though he knew Cos's profile well enough to understand he was a reciprocal to her personality, he was encouraged by their few similarities. They both had a period of $2\pi$, and they both felt positive in the Q1 and Q4 districts and negative in the Q2 and Q3 districts. *Is that why I feel secondary to her? Because I am her reciprocal?*

Cos wasn't the typical superficial girlfriend he usually felt superior to. But he was sure he could make her attracted to him.

*Why couldn't I smile at her or go up to her and talk with her at the 0° gate? Other people are so easy to go up to and shoot the bull with. I don't even admire them, but Cos drives me crazy. I wish Sin were out of the way and that girl Tan. What is it about her? I wouldn't go up and talk with her either. Why? What is it about these people? There's a personal power about them. Why don't they see my power? They are probably just jealous and intimidated. Like most people. Csc is easy because she's weak. She doesn't have any power that radiates from her.*

● ● ●

Csc was sitting in her cybersecurity class in the Q2 district when she received a text from Sec:

> Can we meet up again at the café? I'm really enjoying studying with you.

Csc smiled to herself, not realizing that a blush had turned her cheeks and forehead a bright red. Her fingers were shaking a little as she sent a text back to him:

> Sure. I'm in class in the Q2 district near the café. I can meet you there after class.

Sec texted back,

> See you then.

Csc ran her fingers over Sec's text: "See you then."

## CHAPTER 14

• • •

Csc was headed for the café door when she noticed Sec leaning against one of the café windows, looking somewhat annoyed. *Is he mad at me? I'm not late.*

"Hi. Are you OK?" Csc asked.

Sec looked at her and grumbled, "Does he ever leave this café?"

"Who?"

"Sin. He's always here."

"A lot of people study here regularly. You can't eat or drink in the library. Why does he get on your nerves?" Csc asked. She couldn't see why anyone would think badly of Sin, although she did understand jealousy.

"Don't worry about it," Sec said. "Let's get a table."

Csc ordered her drink at the counter. She looked over at Sin, feeling secondary to him as always, but also grateful to him. She remembered how well he had treated her at the 180° gate. She didn't look away this time when he looked up from his computer. She smiled at him, and he waved.

As Sec and Csc sat down at a table some distance away from Sin, Sec asked her, "So have you seen Cos lately?"

"Sure. I saw her by the clock tower the other day. Why?" Csc asked.

"One of the guys on the team was interested in her. Should I encourage him to ask her out, or is she seeing someone already?"

"I don't know. I see her with a lot of people. The last time I talked with her about a guy, she was asking about Sin. It was on the day you ran into me and we got something to eat here for the first time together. Cos was helping Tan move in with her that day. Anyway, why would you encourage your teammate if you want to go out with her?"

Sec knew he needed to control himself and not let Csc see he was seething at this news. "My teammate can just contact her on Unit Match if he wants. I'm not worried about him. Cos would never choose him over me, would she?"

"I can't imagine why anyone would choose anyone over you," Csc said mechanically.

"Do you have a profile on Unit Match?" Sec asked.

"No. Do you?"

"Do you think I need one?" Sec asked arrogantly.

"I guess not." The more Csc tried to appear not to care when she was stung by words, the more she felt emotion surfacing, which increased the risk of others being able to see.

*Good. She's jealous.* "No matter how many girls I've been out with, not one is as smart as you are," Sec said skillfully, keeping her on a leash. He loved it when Csc blushed.

"Why don't you put your profile on Unit Match?" Sec suggested. "You can get into conversations with other people and learn all kinds of things about our fellow students to share with me. As you can tell, I'm looking for an excuse to keep studying together," Sec said with a wink.

"Sounds like a plan," Csc said, trying to sound casual.

"Hey, you know that favor you did for me the first time we were at the café together? How about doing it again? I'll wait for you outside."

While waiting for Csc, Sec's thoughts turned to Sin. He now had another reason to loathe him, and he was becoming impatient. *Where is she?* The cold was starting to bite. *If Sin were out of the way, I would have a chance with Cos.*

Csc finally emerged from the café as Sec walked past her in a nervous rush, as if he had forgotten he was waiting for her.

"Hey!" Csc called.

Sec stopped suddenly and turned back around to see Csc standing by the café door with a questioning look.

"What gives?" Csc asked.

"Just come on," Sec said and started walking again. "Did you do it?"

"Yes. I took the Wi-Fi out again. Couldn't you hear the entire café moan? Sin wasn't happy either. What got you all riled up?" Csc asked as she followed Sec.

"It doesn't matter." After a minute of walking in an awkward silence toward nowhere, it seemed, Sec stopped again. Looking out at nothing in particular, he said, "I don't know what to make of Sin."

"So? What's the issue?"

"He never really says anything, but he looks at you like he can see right through you. I don't trust him."

"Of course you don't, if Cos is interested in him," Csc smirked under her breath.

"What?" Sec looked at Csc with a sharp, impatient look.

"Forget it!"

Sec put a hand on Csc's shoulder. "You're forgetting you're becoming my best girl, aren't you?" he said in a quiet voice.

"By the way, are you forgetting that you asked me to find out who sent the Invitation to Meet without a name or location that Cos was interested in?" Csc tried not to glare at him.

"No, I haven't forgotten. I figured it would take you some time," Sec said.

"Ha! Well, it was Sin!"

CHAPTER 15

● ● ●

SIN WAS OUTSIDE OF THE library at $7\pi/6$, thinking about Cos's message to him. He was sitting on one of several stone benches lining the walkway to the library entrance. The library didn't open until ten o'clock on Saturdays. He didn't know whether Cos was interested in his profile or just wanted to manage someone's missing information.

Coincidentally, Cos was walking toward the library entrance at that moment. As she was coming down the walkway, she looked up at the sky with an appreciative gaze. He liked her serene countenance. She seemed very even tempered. He could see she was looking at something specific. When she saw him, she paused for a moment.

"Hi, Sin..." She turned to walk toward him. He thought she seemed both surprised and glad to see him.

"Cos. I guess we both had the same plan today to study at the library. We're just a bit early. Have a seat." Sin moved to one side of the bench to make room for her.

Cos sat down and looked toward the clock tower. "Ten more minutes. I forgot the library opens later on Saturdays. I don't study here very often. Just once in a while with my roommate, Tan. She loves it here. There's something about this district I don't care for. I'm not sure why, though; it's hard to verbalize."

Sin watched her confident but unassuming manner as she spoke to him. Her bright face triggered some emotion in him he couldn't define just yet. He liked it. "Interesting. This is not my favorite district either, but the library has some reference materials I can't get online."

After a short pause, Sin asked her, "What were you looking at in the sky just now?"

"Oh, the moon, actually," Cos said, pleasantly surprised at the question.

"The moon?" He looked up to see a translucent white third-quarter moon against a blue sky.

"I'm an astronomy major. I like to keep track of the phases of the moon each month. I love how predictable its cycle is. I'm drawn to sinusoidal patterns."

"Me too" was all Sin could get out for a moment. Had he ever met someone who viewed the world as he did? He hadn't expected to be affected in this way after just five minutes of talking with someone.

They both looked at each other, not knowing what to say next, even though they both felt as if they had a lot to say.

"Well, it looks like it's time to hit the books," Sin said, standing up as the library doors were being unlocked. "Hopefully, we can talk again soon."

"I'd like that. Study hard." Cos stood up and smiled at him. They both walked into the library and went their separate ways.

● ● ●

*Wow, what a contrast,* Cos thought. She thought about her repulsion to Sec, remembering his leaning on the 0° gate looking at her, and comparing that feeling to how she had just felt talking with Sin. Would he ask her out? She hoped so. Hopefully, it wouldn't be too long before they met and talked again.

Cos grabbed the table at which Tan normally studied. Tan wanted to run an errand before going to the library, so Cos had left the apartment later. As she powered up her laptop, she thought she would check Unit

Match before hitting the books. She was anxious to see whether Sin had a profile on the matchmaking site.

Not seeing a profile for him, she was about to log out of the site, but the thought came to her to view the mysterious profile again—the one without a name or a meeting place. Both the profile and the Invitation to Meet had not changed. They were still incomplete. Cos wasn't sure why this bothered her—only that their profiles were so similar she really wanted to find out who this person was. She wanted to meet him, but that feeling had faded somewhat since her conversation with Sin.

"Hey." Tan put her books on the table and sat down. Her smooth hair was up in a messy bun, and she looked awake and ready for the day. That was her usual mood, especially when she was in the library—her favorite place on campus. Cos marveled at how Tan was drawn to the environment of the building, while Cos wouldn't be coming to the library if Tan didn't spend time there. She liked to study with Tan.

"Hi, Tan. I just got here a few minutes ago."

"Still interested in the loony ad?" Tan joked as she looked at Cos's computer screen. "Oh, the meeting is on the night of our department awards banquet. Sorry, I won't be able to join you."

"Ha. Yes, I'm still interested. Can you believe he left out information on purpose so that anyone who was interested would need to figure it out? Isn't that a little self-important?"

"Yes…but what I want to know is why you can't leave this profile alone if he appears so conceited," Tan said, somehow without any sting of criticism.

"I know," Cos sighed.

"Is there any information in that hyperlink?"

"What hyperl—" Cos realized she had overlooked the hyperlink under the Invitation to Meet. It also registered with her that there was a hyperlink included with his reply to her message.

Cos opened the message from the mystery person. It was the same hyperlink that was under the Invitation to Meet. Clicking on it, Cos saw a dated story about a man applying for a job at a telegraph office. Cos and Tan read the story together. They looked at each other and smiled with amusement. No longer perceiving this guy as self-important, they accepted the appealing challenge of finding out who he was.

CHAPTER 16

● ● ●

Two tickets were tossed on the homework papers in front of her that she was working on in the café. Csc picked them up and looked up to see Sec.

"Tickets for the championship game," Sec said proudly. "They're a peace offering. Sorry I barked at you the other day. You know more than anyone that Sin gets under my skin. I took it out on you. I mean it when I say you're my best girl. Spending all this time together, I don't think I could do without you anymore."

"Thanks," Csc said.

Sec sat down at the table with Csc. "You can take anyone you want with you. Third-row seats," Sec said.

*Who could I take? No one would go with me unless they were using me for the ticket.*

As if he had just read her mind, Sec said, "You could see if Cos would want to go with you. These are sought-after tickets. Not easily obtained. Maybe if she could see me play and hear the crowd roaring for me, she would warm up to me and see I'm a great catch." Sec grinned.

"I'll see what I can do," Csc said.

"I wouldn't want anyone else helping me except you, Csc. You're smart and my favorite friend to spend time with," Sec said, hoping he would see her blush. He did. *That was easy.*

"I'd do anything for you, Sec," Csc said.

Feeling accomplished, Sec sat back in his chair. "So have you told Cos that Sin is behind the mysterious Invitation to Meet with the missing name and location?"

"No," Csc said.

"Good."

Csc looked at him expectantly.

"I want you to create a meeting time, date, and location and insert it into the mysterious Invitation to Meet," Sec said.

"What?" Csc said. "Hacking into someone's account and viewing his information is one thing. Now you're asking me to tamper with the information Sin posted on his Unit Match account."

"Can you do it?" Sec asked.

"You'll need to make sure you have an updated profile on Unit Match so I can use it for the Invitation to Meet," Csc said.

"Use Sin's profile. Don't change it. Just insert the meeting date, time, and place."

"Why do you want me to do this?" Csc asked.

"You'll see."

CHAPTER 17

• • •

COS CONSIDERED ASKING SIN WHETHER he would be interested in putting his profile on Unit Match. She was walking counterclockwise around the campus to class, thumbing through Unit Match on her phone. Halting abruptly, she sat on a nearby bench. The mysterious Invitation to Meet was no longer mysterious. Information previously absent was now present and staring back at her:

| Invitation to Meet | |
|---|---|
| Date | February 26, 2017 |
| Time | 12:00 p.m. |
| Place | 0° gate |

The identity of the person was still missing, but with a date, time, and place, she would find out anyway. She continued on to class with excited anticipation and began to entertain daydreams that the person whom she would meet would be Sin.

Since Sin did not have a profile on Unit Match that she knew of, she couldn't view it to determine whether they would be a good fit; this was the reason she wanted to encourage Sin to put his profile on Unit Match. However, even with the absence of computer verification that she and Sin were compatible, intuition and feeling trumped the qualitative evidence his profile was sure to deliver.

*Have I ever felt connected to anyone like I feel connected to Sin? Is this what it feels like to be kindred spirits with someone, somebody similar to you in character?* The brief moments they had talked with each other had bonded them, in her view, in a deep way she had never experienced with the many acquaintances on campus she had known for what seemed a lifetime.

The profile of the person whom she now had a plan to meet was almost an exact match to her own profile.

*Will this prove to be a kindred spirit as well? Will a well-matched profile also satisfy the depth of connection that people long for between each other but can't explain?*

The thoughts swirling around in Cos's mind made it challenging for her to focus on the lecture material in class. She really looked forward to February 26!

● ● ●

Cos reached for her phone to turn off her alarm without opening her eyes. She lay in the dark for a few moments, smiling peacefully. Today was the day. The meeting was at the main entrance, at the 0° gate, at noon.

She would get to meet her match, according to Unit Match. Sin's face flashed in her mind again, but she pushed it out. *I don't want to be disappointed if Sin is not the one I meet. I need to keep an open mind. Are there other people also interested who will be at the meeting place? That might be awkward, but that's how it works.*

*The Trigonometry Tryst*

"Do you want me to come with you to the tryst today?" Tan joked with a wink when she saw Cos come out of her room for a shower. They both giggled.

"That's OK, but thanks. Besides, you'll still be in class at noon. I'll be so excited to tell you about it, though. I'll come over to the library afterward."

"OK. See you then. Good luck!" Tan said as she left the apartment for the day.

When Cos left the apartment, she looked up at the sky, wondering what she would see this morning. Today was the day of the new moon in February. A brand-new cycle would begin again today for the following month, the moon repeating its periodic orbital path around the earth as it had done for eons. Even though no portion of the moon was illuminated at this time, she could still see the faint outline of the moon, as though it were behind a translucent partition the color of the sky.

● ● ●

Cos sat on a bench at the 0° gate five minutes before noon. She admirably lacked a calculating aspect of character. There was no need to plan to be fashionably late or manipulate whose court the ball should be in. There was no shame in sincerity. As Cos took a deep breath and waited with a confident hope, she looked around her. Her eye caught the circumscribed insignia at the crest of the black iron arc that bridged the main entrance to the university.

*How fitting that the date and location of this meeting should coincide with the new moon for this month and that I am sitting at the main entrance to the university, with the new moon insignia.* The insignia at this location, circumscribed by the words "Unit University," was a smooth, circular face of black iron. Cos found meaning in this symbol because of her passion for astronomy and periodic cycles. She thought of the other three gates, each having a portion of its circumscribed insignia illuminated by a silvery sheen. No silver sheen at this gate. A fresh beginning. Standard position.

Cos was startled out of her thoughts by the shrieks of a group of girls gathered at the gate. They were all looking and pointing at a guy who had just approached them.

"Is it you?" several of the girls cried, hysterically jumping up and down.

Her view obstructed by the groupies, Cos stood up to see the guy who was evidently behind the Invitation to Meet and the profile she was so attracted to—the profile that matched her own.

Cos felt a knot form in her stomach as she stared at the guy who was her supposed match staring back at her. He stood in the midst of the group of girls, all of whom were trying to grope his shoulders, arms, and chest—all seeming to say, "Pick me!" But Cos was the one Sec was watching intently.

She had prepared herself to keep an open mind and not feel disappointment if Sin was not her match. Far from being mildly disappointed, Cos felt a drain in her body. She felt light headed, and her heart seemed to empty as the hope dissipated. *OK, so he's not Sin. But Sec? How could* he *be my match? Why is he looking at me?*

Sec had a group of girls smothering him; he was a star athlete on campus. He could have had his pick of any of those girls who *wanted* to go out with him to boost their social reputations. Cos turned to walk away from the group quietly, feeling the same repulsion she felt whenever she was near Sec.

Feeling a hand on her shoulder, Cos turned to face Sec.

"So you were interested in my Invitation to Meet," Sec said with a charming smile.

"I…" Cos couldn't think of anything tactful to say to smooth over this encounter.

"I'm glad," Sec said quickly. "I'd like to take you out this weekend. Friday or Saturday night?"

Cos was aware of the girls looking at her. She smiled at them, not triumphantly but just to be nice. "There are a lot of girls wanting to go out with you. They'd be so glad if you chose one of them. I have things I need to do this weekend. Thank you for asking, though." Cos turned again to leave.

"Hey!" Sec caught up to Cos again. "Why did you show up? I thought you were interested. What's going on?"

Chills ran through Cos's body as she saw the look in Sec's eyes.

"Do you know what a chance this is that you are giving up? Those girls would give anything to go on a date with me," Sec said in a low but aggressive tone.

Cos realized Sec had a grip on her arm. "Then I would ask one of those girls instead."

Responding to Cos's determined look, Sec released his grip on her arm and watched her turn to walk away again. *This isn't over.* He turned back to the group of girls to charm them, ensuring they would continue to idolize him.

CHAPTER 18

● ● ●

COS WALKED WITH DETERMINATION CLOCKWISE around the campus to find Tan at the library in the Q3 district. *This doesn't make any sense. His profile is almost an exact match to mine except for our definitions. If his profile is a match to my own, why do I feel like a reciprocal match to him instead? Something doesn't add up. I don't trust him. There's no doubting that. I suppose he could have lied on his profile. There is no way we have the same domain and range in our personalities. Our definitions have got to be exact reciprocals. No way. He lied.*

Thoughts swirled around in Cos's mind. Her even temper was being tested. What did she feel? Anger? Despair? She was used to feeling contentment and

*The Trigonometry Tryst*

hope, always trying to help others feel better and see the bright side, always looking for the good in others. When she didn't see it, she had the diplomatic skills not to let it show. She was good at not giving herself away. Others were comfortable around her; however, she couldn't hide her feelings at this moment. Her chaotic and distressed thoughts made her unaware of anyone and anything around her.

"Cos? Hi."

Cos stopped abruptly when she heard her name. "Sin."

Sin thought Cos looked almost frightened. Her red eyes looked startled. *Has she been crying?* "Are you OK, Cos?"

Cos took a deep breath and let her shoulders fall. "I'm so glad to see you, Sin."

"Do you have a minute? Come on. Let's sit down." Sin motioned to the bench on the circumference of the campus at $5\pi/4$, the address of the proposed building design, Tan's senior project, in the Q3 district, which they both agreed they came to only for the library.

Cos felt such relief at hearing Sin's voice and being in his presence that it took a lot of effort to keep herself composed, not to let her feelings spill out in an emotional tidal wave. She didn't want to scare him away.

They both sat down on the bench, and neither one said anything immediately. The time on the clock tower could be clearly viewed from where they sat. It was 12:25 p.m. Cos liked the steady reliance of the clock hands. The periodic behavior of the clock reminded her of the periodic behavior of the moon's orbital path around the earth. *A new moon occurred today. The beginning of a new lunar period occurring every twenty-nine days…hmm.* The meeting with Sec had been scheduled for noon. *The beginning of a new period on the clock. The meeting place was at the main gate of the campus, 0° and standard position. The beginning of a new period if you started walking from that point and made a full circle around the entire campus…*

"What's on your mind?" Sin asked after a quiet moment.

Cos turned to look at him. "Just a disappointment. That's all. I don't know why it pushed my buttons so much."

Sin looked at her expectantly but patiently, so she continued.

"Do you get on Unit Match much?" Cos asked.

"I like to browse it when I get a chance," Sin answered casually.

*Then why can't I find your profile?* Cos decided to wait to ask him whether he had a profile. "It's just I saw a profile that was almost a perfect match to my own. This person posted an Invitation to Meet with a complete profile filled out except for his name. The Invitation said for anyone interested in his profile to meet at noon today at the main entrance gate. I decided to go, and I was looking forward to it. When I saw who this person was, I was disgusted."

Cos looked away from Sin and continued. "I don't know this person well, but when I've been around him, I've just wanted to get away as soon as possible." She looked back at Sin. "You caught me at just the right time. I left the meeting as soon as I could. Loads of girls were there, crawling all over him. Have you ever had an experience like this?" Cos loved the sincere look on Sin's face. She knew he listened with genuine interest.

"I can't say I've had that experience from an online matchmaking site. It's interesting that your profiles were almost an exact match," Sin observed.

"That's really what's bothering me, I suppose," Cos confessed. "I had more confidence in Unit Match's matchmaking algorithms than I should have. I couldn't have met a person more reciprocal to me in personality and character. Do you think he lied? I've heard plenty of stories of people not turning out to be what they claimed to be on the Internet."

"It's possible. People are so complex, though. I can't believe that a short personality profile can include a human being's entire character. A personality profile is a good filter to start with, but it's only a start," Sin reflected.

"You're right. It's a good filter. How you feel and what you sense about a person when you are with him or her is the true indicator, isn't it? I feel so silly for reacting the way I did. I'm glad you were here. I feel much better." Cos let out a deep breath and smiled at him. "Thanks."

Sin noticed that the frightened look in her eyes and the distress in her face had dissipated. "Sure. Anytime," he said softly. "Are you headed to the library?"

"Yes. I'm meeting my friend Tan there."

"Well, until the next time I run into you again." Sin winked.

Cos smiled at him and continued on to the library. She could have talked with Sin for hours. She loved his calm demeanor. She sensed wisdom and a depth about him that couldn't have been uncovered in a personality profile on Unit Match. *Ask me out, Sin!*

CHAPTER 19

● ● ●

SIN FOUND A TABLE AT the café in the Q2 district and booted up his laptop. He couldn't get his mind off the experience Cos had related to him. As interesting as he found it that someone else had posted a profile and an Invitation to Meet with missing information, he felt uneasy about Cos's encounter with the guy. Their profiles were almost an exact match, yet she was disgusted with him.

Sin pulled up his own profile and Invitation to Meet on the computer to put his mind at ease; however, he quickly realized his suspicion was validated, and he leaned back in his chair to absorb what he was looking at on the computer screen. His own account had been tampered with. *I don't believe this.* Sin ran his hand through his hair as he took a deep breath. He could see that his profile remained the same, even with the missing name, but the meeting time, date, and location had all been altered—altered to match the time, date, and location of Cos's meeting with the guy she couldn't wait to run away from.

● ● ●

Cos couldn't wait to talk with her friend. She found Tan in her usual study area in the library and sat down with such swiftness that Tan immediately stopped what she was working on and turned to Cos with a questioning look.

"You won't believe this. The guy that showed up at the meeting place was Sec."

"Sec! What are the odds? That was Sec's profile?" Tan asked incredulously, trying to convince herself she had heard Cos correctly. The two girls looked at each other for a moment.

"I'm so sorry," Tan said. "I know how you feel about Sec, and you've been looking forward to meeting the person behind this Invitation for a while."

"I shouldn't be making a big deal out of it. These experiences come with the territory when you jump into online matchmaking. I get that. It's just that his profile was almost an exact match to my own. If Unit Match calculated Sec to be my match, there is no reason to spend any more time on this website," Cos reasoned.

"Maybe Sec lied on his personality questionnaire," Tan said.

"I hope so. I wouldn't be confused anymore, if that were the case." Cos and Tan started to laugh.

"Wait. Pull up the Invitation," Tan said. "Let's look at Sec's profile again."

Cos hadn't booted up her laptop yet, so they used Tan's computer to get on the Internet.

"What?" Cos's mouth dropped open. She and Tan were both looking at the original Invitation to Meet with missing information:

| Invitation to Meet | |
|---|---|
| Date | April 30, 2017 |
| Time | 8:07 p.m. |
| Place | |

"Are you sure you clicked on the right Invitation when you saw the information for today's meeting?" Tan asked.

"I'm sure. Plus, the profile is exactly the same," Cos asserted.

"The hyperlink is still there to that story about the telegraph operator," Tan said. "The person behind this Invitation wanted you to figure out the location of the meeting. Then, all of a sudden, he changes the time and date of the meeting and reveals a meeting location.

You just came from that meeting. Now the meeting is reversed back to the original time and date and the original intent of wanting you to figure out the location."

"What's the point now of figuring out the location of this meeting?" Cos asked. "We already know Sec is the person behind this Invitation. I don't want to have another encounter with Sec. I've lost all interest in this Invitation and profile." Cos sat back in her chair and folded her arms.

As much as Tan had previously tried to convince Cos not to take the Invitation seriously, now she was the one whose curiosity was captured—not because she was attracted to the profile, but because something just didn't add up.

"You're right about doubting that this profile could be Sec's," Tan said. "I don't know him that well, but I could see where his period is $2\pi$; however, there is no way this is his domain and range, like it is yours. His definition would be the reciprocal of your definition.

Either he lied, or he impersonated somebody at the meeting today."

"I just got the chills when you said that," Cos said, unconsciously hugging herself.

"I don't know what's going on," Tan said, "but what if you did try to figure out the location for this meeting on April thirtieth at 8:07 p.m.? It looks to me like there is a mystery to be solved that is not all about Sec. Maybe you'll meet the real person behind this profile you are so attracted to after all." Tan looked from the computer screen to Cos and raised her eyebrows. She was glad to see a smile start to spread across Cos's face.

CHAPTER 20

● ● ●

Csc's eyes followed Sin as he crossed the café and sat down at a table. From her seat at her own table, she had an unobscured view of Sin, although she wanted to watch him without being easily viewed herself. She waited, ready to look away if he looked up in her direction, and watched his face as he realized his Invitation to Meet had been altered.

Sin leaned back in his chair and ran his hand through his hair. He still looked composed—not stony, but in control of himself. *What a contrast. He's probably already thinking about what he's going to do. Making plans. Using his intelligence to solve his problems. Bypassing the emotional black hole of fear, anger, loneliness, and insecurity I would be spiraling down if this were happening to me. I would not be capable of making plans. How does he do it?*

Guilt was starting to invade Csc's thoughts, until she pushed it out of her mind. *He's not me. He hasn't had to deal with the same things I have. He has enough admirers to last him a lifetime. He can handle it. I need a friend. I can't lose Sec now.*

Csc's thoughts began to rewind to the image of Cos's face at the meeting. From her obscured position, Csc had been close enough to see Cos's facial expressions. She had seen Cos's expression of anticipation as she stood up to look over the group of girls to see who the mystery guy was. Csc hadn't taken her eyes off Cos as she observed Cos's expression gradually turn from anticipation to confusion and hurt as she realized this wasn't the moment she had hoped for.

Csc had smiled inwardly when she witnessed Cos's setback. She had watched the entire encounter between

Cos and Sec and knew Cos had walked through the gate trying to conceal the distress on her face.

The image of the hurt on Cos's face replayed in Csc's mind. She sat at her table in the café, her awareness shifting from Sin to Cos. Csc was aware of her jealousy. *Why am I happy Cos didn't get what she wanted? Leave it to Cos to think she is too good for Sec. Any girl would die to be singled out by Sec. He chose her among all the girls, and she walked away from him with an attitude! She has no idea what it feels like to be friendless.*

Yet competing thoughts were elbowing their way into Csc's mind. *Sin and Cos have never hurt me. They've always been kind to me, but I don't care that I've hurt them. Why?* As Csc made her choice between which thoughts would take precedence in her mind, she started to feel dejected. *Their hurts don't come close to my hurts. They will be socially successful no matter what happens. They will always have friends coming to their aid. They are not desperate. I won't give Sec up.*

"You look lost in thought," Sin said.

Startled, Csc looked up at Sin's compassionate smile. She saw sincerity in his eyes. "I guess I have a lot on my mind."

"Well, I know you'll get it figured out. Have a good evening."

"Thanks. You too," Csc said. Sin left the café for the day. Csc felt nervous and confused.

CHAPTER 21

• • •

INSTEAD OF WAITING TO RUN into Cos again, Sin decided to seek her out. He wanted to know the identity of the mystery guy who appeared to be responsible for hacking into his account on Unit Match. He hoped she would still be at the library studying with Tan. He walked counterclockwise from the café in the Q2 district to the library in the Q3 district.

As disturbed as Sin was that his account on Unit Match had been hacked and his posts had been intentionally modified, he was also pleased that Cos was attracted to his profile and wanted to meet him. Walking with his usual focus and his hands in the pockets of his brown leather jacket, he smiled internally. *Will Cos give up, or will she persevere and figure out the real tryst I originally posted by hearing the message in the telegraph machine telling her where to meet? I hope so. I like being with Cos.*

Sin walked into the library and was relieved to see Cos and Tan sitting at a table. He walked over to them and observed that they didn't appear to be studying but were quietly talking with each other.

"Hi again," Sin said to Cos. He then acknowledged Tan.

"Oh, hi!" Cos said with a bright face. "Do you know my friend Tan?"

"We've met before. It's nice to see you again," Sin said warmly as he shook Tan's hand.

"It's nice to see you too. Have a seat. You are welcome to study with us," Tan said.

"I'd like that, but I only have a minute." Sin took a seat across the table from Cos.

"I was just telling Tan about the surprise at the meeting at noon today," Cos said.

"I've been thinking about it too," Sin said.

"You have? What, specifically?" Cos asked.

"Just that your profile was so similar to the mystery guy's profile, but you said you felt he was your reciprocal," Sin said. "What do you think, Tan?"

"I think Sec lied on his profile. First impressions may not always be correct, but I think Cos should trust her intuition on this one," Tan said in her usual straightforward manner.

"Sounds like wise advice," Sin observed. *So the impersonator was Sec.* "It's interesting that Sec would feel he needed to lie. Wouldn't he attract the people right for him by being himself? But then, I can be an idealist."

"People will lie if they don't like themselves or if they feel insecure about who they are," Tan said, turning to Cos. "Sec probably thought he'd never get a chance with you if he were honest with his profile."

"How did he even know I would be there?" Cos asked. "Besides, it's so foolish to pretend anyway, as you can't hide who you've become for long. Why doesn't he just date one of the groupies? There were plenty of them there who would sell their souls to go out with him."

"Like attracts like, although people can certainly complement each other," Sin said thoughtfully.

"Well, in this case, reciprocals do not attract. That's for sure," Cos said.

"Relationships aren't so black and white, are they?" Sin asked. "So what do you two usually do for lunch?"

"It varies depending on the schedule," Cos said, getting her hopes up.

"I usually eat at the café while I'm studying. It would be nice to have some company. Would both of you like to join me tomorrow?" Sin asked.

"I can't tomorrow, but I'll join you two another time," Tan said. She lightly kicked Cos under the table.

"I can meet you at noon," Cos said.

"Perfect. I'll just be getting out of class in the Q3 district, so I'll grab the bus and meet you then," Sin said and got up to leave.

Tan looked at Cos, who couldn't hide her grin any longer.

"I know you declined lunch so I could have him to myself, Tan. You could have come," Cos said.

"I know," Tan said. "Just tell me all about it afterward."

CHAPTER 22

• • •

"Do you mind if I sit with you for a little while?" Sin asked Csc as he pointed to an empty chair at her table in the café.

"Oh, sure," Csc said, surprised.

Sin set his laptop on the table across from Csc's laptop and started booting it up.

"Would you like something to drink?" asked Sin.

"Thanks, but I'm fine. I'm almost finished with the one I have," Csc said.

"I think you like this café as much as I do," Sin said politely as he sat back and casually placed one ankle on the knee of his other leg.

"As a computer engineering and IT major, I do spend a lot of time on the computer," Csc said. She knew her voice sounded weak and nervous. She couldn't look at Sin for more than a couple of seconds at a time, even though he looked at her intently.

"Speaking of being an IT major…" Sin let his voice trail off and watched Csc go very still. "My account on Unit Match was hacked."

"Really?"

"How do you think that could have happened? Surely the university's firewalls are up to standards," Sin suggested patiently.

"How do you know your account was hacked?" Csc asked.

"Well, I posted an Invitation to Mcct on Unit Match, and someone altered it."

"Huh. Have you given your password to anyone?" Csc asked.

"No. Not a one," Sin said, still looking at her steadily.

"How was the Invitation modified?" Csc stalled.

"Someone modified the time, date, and location of the Invitation while using my profile. The interesting thing is that the meeting occurred with the modified information. You'll never guess who posed as the host of this meeting. It's someone you know," Sin said carefully.

"Who?" Csc asked.

"Sec."

"Sec? Did you see him? Were you there?" Csc asked.

"No, but I was informed," Sin said. "I wonder how he did it and why he would do it."

"I don't know," Csc said. She looked over to see Sec coming through the door of the café. He didn't look happy when he saw Csc and Sin talking together.

"Csc, I'm crushed. You found a new study partner!" Sec said, holding his gaze on Sin.

Sin stood up. "I was just leaving. You're welcome to have my seat," he said as he picked up his laptop to carry it over to another table. Sin met Sec's gaze with a look that kept Sec from saying anything further.

"What was that about?" Sec asked Csc impatiently as he sat in the chair Sin had vacated.

"Sin knows his Unit Match account was hacked. He knows you impersonated him at the Invitation to Meet," Csc informed him.

"How did he find out? Was he at the meeting?" Sec asked.

"He says he wasn't."

"I hope I don't find out you told him," Sec said, testing Csc.

"Why would I do that?" Csc rolled her eyes.

"Well, he can't prove anything. I can always show that it was a coincidence. I'll post the Invitation to Meet with my own profile. Can you backdate it for me?"

"Sure."

"I can always count on you," Sec said, winking at Csc. He noticed she wasn't blushing anymore in response to his flattery. He knew he still had her, though. They had become comfortable partners.

"What is he doing?" Sec asked Csc, who had a clear line of sight to Sin.

"Looking at the back of your head," Csc said.

CHAPTER 23

• • •

COT SAT DOWN IN HIS closet-like grad-student office to e-mail Tan the final survey report for her architecture project. *She'll be happy now.* He attached the file and, trying to send a quick note with it, found himself unable to type. *What do I say?* He was still hurt by Tan's cutting remarks.

He admired Tan and assumed she would settle into an easygoing relationship with him, as everyone did. Cot was easygoing himself and always able to smooth over conflict by ignoring it and not making it bigger. Everyone fell for his approach, so he remained safe. Cot shoved his faults and weaknesses under the rug, never to dwell on them or improve them—never to take responsibility for them. They didn't matter. *Tan didn't respond to my effort to make light of the situation.* He wasn't used to people responding to him the way she had responded. *Those piercing green eyes. She could see through me.*

CHAPTER 24

● ● ●

Cos looked at the mysterious profile again, this time from a different point of view. After rereading "The Telegraph," she wondered what she could glean from the profile to help her determine the location of the meeting. If she could do that, the identity of the person would take care of itself.

For starters, Cos concluded the location would be in either the Q1 or Q2 district. *It makes sense to me that the location would be set in a place you like and feel positive in.*

Back in their apartment for the evening, Cos took a break and watched a sitcom episode while her laundry was drying in the apartment complex's communal laundry facility. Tan thought she would join Cos for a much-needed break as well. She sank onto the opposite end of the sofa from Cos and tucked her feet under her. Cos let out a tired laugh at a line from the show and looked as if she would fall asleep at any minute. Tan smiled and checked her e-mail on her phone one more time for the day. She wouldn't mind turning in soon. She felt as tired as Cos looked.

As she thumbed through the list of e-mails, nothing looked important enough to open. Mostly advertisements and social media notifications, which she just deleted without opening. Then her eye fell on an e-mail with the subject line "Survey." The sender read "Cotangent."

"Finally!" Tan shrieked and jumped off the sofa. "My survey is ready!"

"I'm glad it's ready, but do you really have the energy to get into it tonight?" Cos asked groggily.

"I just got a second wind. Just like you'll get when the total solar eclipse occurs in August. You aren't just going to sleep through it, are you?" Tan said, disappearing into her bedroom to get her laptop out of her bag.

"Touché," Cos acknowledged.

Booting up her laptop, Tan nestled back into her corner on the sofa and opened the e-mail with the survey attached. Bristling at seeing Cot's name, she read the e-mail quickly:

Tan,
Attached is the report for your survey request. Finished by the end of the week, as promised! I hope to see you in the library again soon. Let me know if you need help with anything.
Cot

"Oh brother," Tan said under her breath as she opened the attachment.

"What?" Cos asked sleepily, turning her head toward Tan.

"Cot wants some credibility, but I won't go into it if you are going to fall asleep on me," Tan teased.

"OK, tell me tomorrow," Cos said with her eyes closed.

Tan read the survey report, relieved to find the lot had a slope less than fifteen degrees, so it was on the flat side, simplifying the design. No need to add complexity, as would be necessary if she were building into a steeper slope. A couple of trees on the lot, Tan knew, would need to be cut down since they were where the building would stand. New landscaping would be implemented. She planned to measure the height of the trees using a clinometer to find the angle of elevation to the tops of the trees. Their heights could then be determined. Understanding these heights was important in determining the direction the trees needed to fall when cut down. If the trees fell in the wrong direction, they could cause damage to the neighboring buildings.

Although Cot wasn't responsible for measuring the heights of the trees for the survey, the angles of elevation to the tops of the trees were included in the report. This didn't soften Tan's heart toward Cot. *He can just say he's sorry. These words go a long way when spoken sincerely. Problems can be solved and things won't fester if people just work it out. Instead, people play all sorts of games to avoid saying these words. Amazing.*

● ● ●

Cos woke up the next morning, walked into the living room, and noticed Tan's things still on the sofa from the night before. A drawing was lying on Tan's laptop. Taking a moment to raise her arms in the air and stretch, she kept her eyes on the drawing. Something about the drawing sparked an idea.

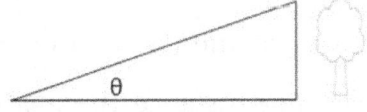

Cos saw Tan's calculation to measure the height of a tree:

$$\tan \theta = \frac{opp}{adj}, \theta = \text{angle of elevation}.$$

Tan walked in the room to see Cos eyeing her drawing. "You like my tree drawing, huh?"

"How late did you stay up last night reading your survey results?" Cos asked.

"Not long after you brought in your laundry and went to bed. If I measured the heights of the trees on the lot, I could plan while I was drifting off to sleep," Tan said.

"Planning would keep me wide awake," Cos said with a yawn. "Looking at your calculation, though, I see you could also have found the height of the tree by $\sin \theta = opp/hyp$." This observation triggered something in Cos's thoughts.

"True, but my calculation is faster because I can easily measure the ground distance, the adjacent side, from the tree to where the surveyor is standing to measure the angle of elevation. If I used the calculation you were thinking of, I would have had two unknown variables: the height of the tree and the hypotenuse of the right triangle," Tan rattled off with confidence.

"You're right," Cos observed. Then it hit her. "That's it!"

"What?" Tan asked, surprised by Cos's sudden reaction.

"The ratio *opp/hyp* looked so familiar to me. How could I have forgotten? It's the mystery person's right-triangle definition on his Unit Match profile! One more clue to trying to figure it out. The author of 'The Telegraph' would be proud of me, don't you think?" Cos said with a shining grin.

"I think so," Tan said, smiling back at Cos.

## CHAPTER 25

● ● ●

TAN WAS STANDING ON THE plot of land for her project, mulling over the results of the survey and envisioning her design in that spot. She badly wanted her design to be selected. Her professor was always proud of her work and thought she stood a good chance. Tan's ambition was good as long as her need to be at the top academically didn't distort her thinking. She knew she had the potential, but she could be so hard on herself, measuring herself against others if she didn't come out on top. As solid as she always appeared to everyone else, she had her own internal issues to deal with.

Visualizing the building against the design for the landscaping gave Tan satisfaction. The feel of creating something in her mind, on paper, using tools and math, and seeing the resulting physical structure was satisfying.

Coming up on the location he had surveyed for Tan's project, Cot saw Tan walking around the plot in deep thought. The only reply he had received from Tan after sending her the survey report was a curt e-mail response to say thanks. *Was she satisfied? Does she know I went beyond the scope of the survey to make her happy?*

"What are you thinking?" Cot asked.

Tan turned to see Cot standing near her with his hands in his pockets and a big grin on his face. "I didn't hear you coming," she said.

"Did the survey work out for you?" Cot asked.

"It's what I needed."

"There were a few additional calculations done for you at no extra cost," Cot asserted with no hint of the big grin dissipating.

"Nice try," Tan said. "You want me to think you've done me a huge favor, huh? I'd say you owed it to me anyway for dropping the ball in the first place. It's good business to try to make it right when you mess up, so the additional calculations were good. Thank you."

Cot wished he could see Tan's bright face, the face he had seen in the library, the face he was attracted to. He couldn't understand why she was being distant and cold to him. He had tried to make up for what had upset her. "You're welcome," Cot said with some irritation.

Tan shot him a look. "What are you irritated at? What did you expect? That I would fall all over you and compliment you on being the sweetest guy I ever met? You're not fooling anyone."

Tan saw the pursed lips and hurt in Cot's eyes. A mutual stare lasted several moments.

"What do you want?" Cot asked with a lingering tone of irritation.

"What do you mean, what do I want?" Tan asked. "I didn't ask for anything. As you can see, I am working on my project."

"What do you want me to do to make you happy? I would like to be friends," Cot said. "Hopefully, this superior attitude of yours doesn't bite very often," he said with a laugh, hoping to make Tan smile.

"What you're saying is that I shouldn't stand up for myself. That I should be sweet and charming regardless of how you want to treat me," Tan said sharply.

"Gee, leave it alone, will you? I'm trying to be nice," Cot griped.

"You're trying to shove a problem under the rug. I'm not an idiot."

"How's that? I made up for it, didn't I? I did those extra calculations." Cot's voice was rising.

"Does that mean you can be trusted now?" Tan asked.

Cot just stared at her.

"Why go through the charades?" Tan asked. "Why not just say you messed up and you're sorry? Working it out builds trust. Trying to gloss over the problem makes things fester and builds resentment. Working it out helps people forget it and move forward. Ever thought of that?"

Her face was so sincere. Cot's thoughts were the complete reciprocal of hers. Diplomacy and political tactics were the way it should be done, not being blunt and direct. Someone was supposed to recognize the tactic and strategy. Tan was supposed to recognize his effort to do the extra calculations as his way of saying he was sorry he had messed up. Political astuteness was how it was done. Cot turned around and walked away.

Tan rolled her eyes and tried to focus again on her project. *I wonder what designs other architecture students are going to propose for the building. It should be efficient, but as my interior designer friend says, "Simple does not mean bare."* Her design would be beautiful as well, with some character added to it.

CHAPTER 26

• • •

Csc didn't know how she was going to invite Cos to the basketball championship game to which Sec had given her tickets as a peace offering. Reaching the gate at 270°, the university bus slowed to a stop. Csc looked back and saw Cos get off the bus. She waited for her to enter the gate with the crowd of students who stepped off the bus at the same time.

"Cos, hi," Csc said to get her attention.

"Hi, Csc. How are you doing today?"

"Fine. I wanted to ask you something," Csc said. She could feel herself getting nervous. She felt Cos would turn her down for sure.

"Sure. What's up?"

As they walked together, Csc pulled the tickets from a textbook she was carrying.

"I have two tickets, third-row seats, for the basketball championship game. I was wondering if you would want to go with me," Csc said.

"Oh. When is it?"

"March seventeenth."

"That's really nice of you, Csc. I'd be happy to go," Cos said.

"Oh, great. Thanks. Here's your ticket. I can just meet you there, or I can come by and get you. You live closer to the stadium than I do," Csc said.

"It sounds great to come by our apartment. You can say hi to Tan when you do," Cos said.

"Sorry, I have just two tickets," Csc said. "Otherwise, Tan could come also."

"No worries. Tan understands these things," Cos said.

*The Trigonometry Tryst*

"So have you been able to find any time to get to know Sin better?" Csc asked.

"What? Oh, that's right. I remember we talked about him the day I helped Tan move out of your apartment," Cos said. "We saw him walk into the café."

"I see him a lot at the café. I study there too," Csc said.

"Actually, he asked if I wanted to meet him for lunch there tomorrow. I'm excited to get to know him better. I just don't like the Q2 district. Tan likes to study in the Q3 district in the library, and I don't like that district either, but I don't notice it so much when I'm meeting friends there," Cos said.

"That's great that you'll get some time with Sin. What time are you meeting him?" Csc asked.

"Noon. His class gets out just before that in the Q3 district, so he'll just take the university bus to get to the café by then," Cos said.

"Well, enjoy lunch tomorrow. I'm happy for you. I need to go this way, so I'll talk to you later," Csc said.

"See ya," Cos said. "Thanks again for inviting me to the game."

"Sure. Thanks for coming," Csc said.

Cos thought about what had just happened. *Wow, Csc invited me to the game. Why me? I couldn't turn her down. I feel sorry for her. She's trying hard to make friends.*

Csc felt accomplished. Cos had agreed to go to the game with her, and she had just received some information that would be important to Sec. She sent Sec a text that read,

> Cos agreed to go to the game with me. FYI, she's meeting Sin at noon tomorrow at the café for lunch.

Sec immediately texted back:

Good work! Keep them from meeting. You'll come up with something. I'm counting on you.

Just then, Csc could hear the unmanned university electric bus pull away from the gate, ready to continue its way around the circumference of the university.

CHAPTER 27

● ● ●

TAN WAS WINDING UP HER senior project. The blueprints were complete, and her cost-estimate spreadsheet was resulting in a project within budget, including an estimate for contingency costs.

She knew she should be feeling exhilarated, but something was gnawing at her gut. Had she missed something? She had taken the time to double-check all her calculations, including measurements, current pricing for her building materials, landscaping, and excavation. She decided her thoroughness was causing some anxiety and decided to submit the project. It was required to be turned in on a thumb drive to the architecture department office.

Tan's academic thoughts were often interrupted these days by her confusion about Cot. *He is so strange. I can see he sincerely wants to be friends and have a relationship. I like his personality, but he goes about it in such a cowardly way when conflict is involved. Why am I trying to make him all good or all bad?*

After submitting her project, Tan took a deep breath, walking by the project site at $5\pi/4$. Daydreaming about her design being selected, she didn't see it at first, but as she turned from her daze, she recognized something she hadn't identified until now. *Something doesn't look right.* Then it hit her. As she took it in, she could feel her gut tighten. *The slope of the plot can't be less than fifteen degrees. It looks steeper than that.* It was obvious to her now. Why hadn't she seen this before? She had relied on the survey report, but as a competent professional would have done, she now discerned the problem.

It was similar to professors encouraging students to do their math problems in longhand until they understand what they are doing. If students plug input values into a calculator or program of some kind that spits out an output value, they will not understand what is going on in the problem and thus will be unable to discern whether the output value or answer is correct or reasonable. Once students understand the concept and method to solving a problem, they can efficiently use coded programs.

Tan decided she would measure the slope herself to get a concrete answer. *Then I've got to get my thumb drive back before the committee looks at it!*

Borrowing a spool of string, a leveler, a tape measure, and a straight post from a lab at the nearby engineering building, Tan fixed the string at the top of the slope and pulled it taut as she walked down the sloping plot until she was at the bottom of the slope. After cutting and tying the string to the post, which she inserted into the soft ground at the bottom of the slope, Tan used the leveler to ensure the string was horizontal. She then used the tape measure to measure the horizontal string and the distance from the ground to the point where the string was tied to the post. Using the formula $\tan\theta$ = *opp/adj*, she determined that $\theta$, or the slope, was definitely greater than fifteen degrees. In fact, it was twenty-five degrees.

*Did Cot measure wrong, or was it a misprint in the report? Did he accidentally type a one instead of a two?* Repercussions that would result from the mistake swirled around in Tan's mind. *I won't be able to use a slab foundation unless I incur a lot of extra excavation costs for the fill needed. I could add a basement instead. Basement rooms could be used as offices or labs, but there's not enough money in the budget. What about a crawl space instead? Drainage will be affected. What am I going to do?*

Tan covered her face with her hands and took a deep breath. She would deal with Cot later. The first thing to do was to get her thumb drive back. She went to the architecture office and asked her professor's secretary for it.

"The professor is looking at it now," the secretary said.

"What? I just turned it in an hour ago," Tan said.

"Well, it's not like he had a backlog. You were the first one to submit a design, and he wanted to view your work. You should take that as a compliment."

"Thanks. Is he in his office?" Tan asked with a little resignation in her voice.

"Yes."

Tan walked down the hall and saw that her professor's office door was partly open. She didn't need to knock, since he saw her immediately.

"Come in, Tan," he said.

*The Trigonometry Tryst*

"Professor, I have a problem," Tan said as she walked in to face him at his desk.

"What can I help you with?"

"I understand that you have the thumb drive with the design I submitted for the university building," Tan said.

"Yes, I do."

"I've just realized I made a mistake in the design that I need to correct. I'd like to have it back to correct the mistake and resubmit the design."

Tan's professor looked at her in contemplation. "How did you conclude there was a mistake in your design?" he asked.

"It's the slope of the plot. The survey reported a slope of fifteen degrees. I walked by it just now and realized that the slope must be steeper than fifteen degrees. I measured it myself, and I measured a slope of twenty-five degrees. I've got to modify the design of the foundation or add additional costs for more extensive excavation and fill," Tan said without beating around the bush.

"Very well," the professor said. He pulled the thumb drive from his computer and gave it to Tan.

"Thank you, Professor," Tan said and left his office feeling embarrassed. Her cheeks felt hot. *Have I just hurt my chances of winning? I wanted to win so badly.*

By the time Tan reached the project site, she was angry as well as embarrassed. She walked looking at the ground and with her hands in her pockets, oblivious to anyone or anything happening around her. *Is all my work for nothing?*

"Tan? Hi."

Tan just stopped cold in her tracks and looked at Cot fiercely. She was unsure how to handle this encounter.

Cot stepped back. "What now? Why are you always angry with me? Are you still upset over the delay of the survey?"

*The Trigonometry Tryst*

"I wish the delay had been the only problem," Tan said icily.

"What do you mean?"

"Rather than doing extra calculations in the survey, I wish you had made sure the required calculations were correct!" Tan charged.

Cot didn't know what to say.

"You measured the slope wrong! It was off by ten degrees. Do you know what that means for my design?" Tan turned around to finally exhale.

Cot stood still to let Tan's comments register. "How do you know the measurement was off by ten degrees?"

"I could tell by just looking at it. Something was wrong, so I measured it myself. I had already turned in my design when I noticed it. Then I had to go retrieve my design from my professor. How do you think that felt? Or do you take nothing seriously?" Tan asked furiously.

"Look, I don't know what happened, but I assure you I didn't make a mistake intentionally," Cot said quietly.

"Mistakes aren't all innocent, Cot. Did you double-check your work? Did you do repeatability measurements? Did you calibrate your instruments? Or were you just lazy? Mistakes can be avoided," Tan pressed.

Cot remembered his state of mind when he was performing the survey. He knew he hadn't been at his best. He remembered being distracted by Tan's reaction to the survey delay. It was windy that day. *Did I take a reading without making sure the tripod was level?*

"I've got to redesign the foundation now. I don't know if I have enough money in the budget or if I have to take out other design elements to compensate for it. Cot! Do you realize that your commitments have an effect on other people? Do you even care? What if someone treated you like this?" Tan was exasperated.

"Tan," Cot said calmly, "I made a mistake. I am sorry. It was not intentional. Let me get my instruments, and

I will help you with anything you need to correct your design and get it resubmitted."

After a minute of silence with both of them looking at each other, Tan said, "Thank you for saying you're sorry and for trying to make it right. I could use your help."

Cot let out a sigh of relief. "I do care, Tan, and I want to be friends with you. Give me a chance to try to measure up to you."

"Are you making fun of me?"

"No, Tan. I'm not making fun of you. I just don't know how to be as ideal as you are. I feel I can't do anything right to make you happy. I always feel secondary to you," Cot said.

"I make mistakes too, Cot. The important thing is that when you do, you own up to it, work it out, and make it right. Like you did just now. Now we can move forward, and you are earning my trust. Do you understand where I am coming from?"

"I understand that is how you see things. I've always felt that smoothing things over and avoiding bluntness is more tactful," Cot said.

"Well, you may avoid arguments, but deep down, trust is broken, even if the person doesn't say anything. If a problem isn't worked out, things fester. It becomes an act requiring effort to be civil going forward instead of a sincere, trustworthy relationship."

Cot extended his hand. "Let's work it out. Let me help you."

They shook hands and went to work.

CHAPTER 28

● ● ●

CSC WAS HOPING SIN WOULD walk through the café door. When he did, he saw her immediately and waved. The circumstances in the café were perfect. It was crowded, and Csc couldn't see an empty table. Sin was surprised when Csc did something out of character. She waved him over to her table.

"Hi, Sin. You can sit here if you want."

"Thanks, Csc. It's more crowded than usual, isn't it?"

"You can set your stuff down. I'll watch it while you place your order," Csc offered.

"Thanks."

When Sin returned, he set his drink down on the table and took out his laptop from his backpack. "You seem in a good mood today."

"I am. I got a good grade on my cybersecurity test," Csc said.

"Great! Though I'm not surprised. That stuff fits in your brain pretty well," Sin congratulated her.

Csc couldn't tell whether the compliment was meant as a cutting remark. *Does he know what I've been doing?*

They could see the top of the university bus above the campus wall pass by outside the café window. This provided the perfect moment Csc needed to bring up the conversation she wanted to have with Sin.

"I remember when you got off the bus at the 180° gate on the last rainy day we had. Do you ride the bus a lot?" Csc asked.

"Not a lot. I like to walk unless the weather is bad. Do you ride it?" Sin asked.

"Only if I can. I can't ride it to the 0° or 180° gate. I am asymptotic there, as you observed when you saw me at the 180° gate. I should have told you at the time. I'm sure I seemed very odd to you," Csc said.

"It's OK. I figured it out shortly after you left," Sin said.

"I think the concept of an unmanned electric bus is fascinating. You're an electrical engineering major, aren't you?"

"Yes."

"Do you know—I mean, I'm a computer engineer, so I understand programming—but is it really completely fail-safe? What if the safety features become disabled?" Csc wondered.

"There is an operator in a control room somewhere on campus who is alerted if a safety feature is disabled or if the bus stops unexpectedly," Sin said.

"Huh, do you know where the control room is?"

"Yes," Sin said. "It's in a small building just inside the 270° gate. One of my classes last semester was invited to see it during the testing phase and to talk with the project engineer in charge of the installation last year."

"Huh, I'd love to see it too."

"I'm sure you can. The project engineer won't be there to talk with you, but I'm sure the operator is familiar enough with everything now to answer most of your questions," Sin said.

"Sweet," Csc said. She hid a wry smile.

CHAPTER 29

● ● ●

Csc lingered outside the bus control-center building until she felt unnoticed. It was time for Sin to take the university bus at the 270° gate to meet Cos at the café for lunch, if all was going according to the way Cos had described their plan.

No signs hung on or around the door prohibiting entrance, so she opened the door and peered in. The operator turned around at the sound of the door opening.

"Can I help you?" the operator asked.

"Sorry. I don't know if I need an appointment. I just wanted to see where the bus was operated from. I'm a computer engineering major, so I'm curious, of course," Csc said.

"Sure. No problem. I can show you around quickly. I can't really give you an in-depth explanation, which you probably want, being a computer engineer. I didn't design it. I just operate it."

"That would be great. Thanks," Csc said.

The operator showed her the controls and the displays he monitored.

"Does the bus have to stop at every gate, or could you keep it running continuously if you wanted to?" Csc asked.

The operator pointed at a button. "All I'd have to do is select this button to keep the bus from stopping, but what would be the point?"

"Sure. There must be safety features that could get disabled, like when you need to do maintenance on the bus?" Csc asked.

*The Trigonometry Tryst*

"Yes. But maintenance is performed with a pair of workers. One worker must be on the bus at all times to prevent a runaway."

"Do things ever go wrong?" Csc asked.

"It never has on my watch," the operator said.

They could both hear the bus approaching the 270° gate and coming to a stop. The operator looked out the window of the building and watched as students got on and off.

"See?" the operator observed, looking out the window. "Runs just as smooth as can be. Always going in a complete circle."

Csc couldn't see whether Sin had gotten on the bus or not. She just had to trust that he had. While the operator was still peering out of the window and had his back turned to her, she quickly pressed the button the operator had showed her to keep the bus running continuously. She was relieved that no alarm or indicator sounded. The bus began to accelerate as normal and disappeared from view.

"Well, thanks for showing me. I appreciate it," Csc said. "Sorry for dropping in on you like this. You've been great."

"Don't mention it," the operator said.

Csc left feeling nervous, although she knew that what she had done wouldn't be a safety issue. The bus just wouldn't stop until the operator was alerted that something was wrong. Sin wouldn't be able to meet Cos. Sec would be happy.

● ● ●

Cos stepped onto the university bus at 270° to meet Sin in time for lunch at the café. She thought she would find Sin getting on the bus here—especially since he had talked about taking the bus at this time at the 270° gate after his class let out in the Q3 district—but she didn't see him. She would ride the bus 90°, get off at the 180° gate, and walk the rest of the way to the café at 5π/6 or 150° in the Q2 district to have lunch with Sin. *I'm so excited!* She could see the clock tower rising above the university wall and recalled her earlier realization that the second hand on the clock tower and the university bus making its way around the circumference of the university both had the same angular velocity. She was surprised to note that the angular velocities of the second hand and the bus were out of sync now; the bus was going faster than its linear velocity of forty miles per hour. The bus was increasing in speed and showed no signs of slowing down as it neared the 180° gate.

● ● ●

The operator in the control room for the electric bus heard the alert when the bus didn't stop at the 180° gate. Frantic but in control of himself, he selected the control to stop the bus. He checked the control board: the button to keep the bus going continuously was lit up. He wondered whether it was an electrical glitch or a safety device had been disabled. He knew he hadn't pressed the button, and he didn't think Csc would have accidentally pressed it when he showed her around just a while ago. He made contact with the university police to check on the people on the bus. He had to remain in the control room.

CHAPTER 30

● ● ●

SIN LOOKED AT HIS WATCH and started to get worried because Cos was late. His class in the Q3 district had ended ten minutes early, so he had decided to walk to the café instead of taking the bus. He and Cos hadn't exchanged phone numbers, so the best he knew to do at the moment was just wait. He hoped she was OK. She didn't seem like the flaky type to him; he didn't think she would forget or stand him up.

Sec came into the café a moment later. He stopped short when he saw Sin. He didn't see Cos anywhere. *What is he doing here? Csc must not have gone through with the plan she came up with and told me about.* He could feel his temper rising.

Over a half hour later, Cos rushed through the café door. She searched for Sin with frantic eyes and looked out of breath. When she caught sight of him, she rushed to his table, not taking any notice of Sec, who was sitting a couple of tables away.

"Sin, I'm so sorry I'm late!" Cos exhaled as she sat down at the table across from him and placed her backpack on an empty chair between them.

"It's OK. Are you all right? Did something happen?" Sin asked, concerned.

"You won't believe this. I got on the university bus at the 270° gate. I ended up having a quick errand to run in the Q3 district, so I thought I would go do that and catch the bus with you. I didn't see you there and figured you had changed your plans," Cos explained.

"I did. Class got out early, so I had time to walk. I like to walk if I can. Anyway, go on," Sin said.

"The bus didn't stop at the 180° gate! It kept going!" Cos exclaimed quietly.

"What?" Sin said.

"Yes. The bus went all the way to the main entrance at the 0° gate before stopping," Cos said. "We got off the bus, and the university police were there waiting for us. They interviewed us, took some notes, and made sure no one was hurt. The bus maintenance crew arrived, and I left. I got here as soon as I could, worried that you were thinking I'd stood you up."

"I didn't think you would have stood me up, honestly. I thought something might have happened. I'm glad you're here now and you're OK. It sounds like the bus will be out of service until it's determined what caused the failure. Let's get some lunch, and you can take a deep breath," Sin said.

Sin went up to the counter to put in their orders. As he waited in line, he looked back over at Cos. Sec was talking with her.

"Hi, Cos. Nice to see you here. Can I get you something to eat?" Sec offered.

"Hi, Sec. I'm with Sin. He'll be back with our lunch in a few minutes. Thanks anyway," Cos said politely.

"Anytime," Sec said. He couldn't think of anything else to say, so he went back to his table.

Sin watched Sec return to his table and started putting two and two together. He had just had a conversation with Csc about safety features on the bus and the possibility of a bus failure. She had been picking his electrical engineering brain regarding the unmanned electric bus. *Why?* Sin reflected back to his conversation with Csc. *I told her my schedule. She knew I was planning to meet Cos for lunch here and catch the bus at the 270° gate just before noon. Sec expected me to be on the bus during the failure. Not Cos. He expected that it would be him who would be having lunch with Cos at this moment.*

Sin brought their lunch to the table. "I saw Sec talking with you. Did he upset you?"

"No," Cos said. "I'm just glad we're having lunch together. I really enjoy your company."

"I enjoy yours too, Cos," Sin said, smiling sincerely.

"I'll be watching Sec at the championship game, though. Csc had two tickets to the game and invited me to go with her," Cos said, taking a bite of her sandwich.

"Csc invited you to go to a game with her?" Sin asked. "I'm sorry. I'm just surprised."

"I know. Csc's peculiar, but I feel some compassion toward her. She tries so hard to fit in, but she's such an odd function. I feel like she'll be offended at anything I say, so I don't say too much. I haven't figured her out. Going to the game with her may be a little awkward, but I'll get to know her better," Cos said.

"Just be careful."

Cos noticed that Sin didn't seem to have much of an appetite today. She stilled herself and looked at him. "What's wrong?"

"I know you feel very uncomfortable around Sec. Did you know he and Csc spend a lot of time together?" Sin asked.

"No, but why should that mean I should be careful?" Cos asked.

"I agree with you. I feel compassion for Csc also. I think she's smart, and she could be a person of influence if she were to become more emotionally mature, but I just don't trust Sec," Sin said. "I just want to make sure you'll be all right."

"Thanks, Sin. It makes me feel good you're looking out for me, but everything will be OK. Sec and Csc aren't dangerous, are they? They're not going to hurt me," Cos said.

"You're right. Maybe I'm overreacting to the bus failure today," Sin said. But he wasn't so sure.

CHAPTER 31

• • •

"Do you mind if I sit for a while?" Sin asked Tan.

"No, of course not," Tan said. "How are you?"

"Well, I'm not just passing by. I specifically wanted to see if I could find you, and I thought I would have the most luck looking here in the library," Sin said.

Tan laughed. "Yes, I'm sure Cos has told you this is my favorite place to study on campus. Cos is here with me sometimes, as you know."

"Actually, I was glad to catch you alone. I'm sure you know Csc invited Cos to go to the basketball championship game with her," Sin said.

"Yes, she told me about that," Tan said.

"Did you think that was unusual in any way?"

"Yes, definitely unusual," Tan said. "It was pretty bold of Csc. She used to be my roommate, you know, and rarely took initiative. She depends heavily on others to take care of her feelings. She's very sensitive. You have to tread lightly around her, but I think it's good of Cos to accept. Cos is like that."

"Yes, she is. Do you remember when I was here last with you? We talked about the Invitation to Meet Cos was interested in, and the host turned out to be Sec?" Sin asked.

"Yes, that was awful," Tan said.

"Well, Sec will be playing at the game, as you know. I'm assuming that Csc received the tickets from Sec. They spend a lot of time together," Sin said.

"They do? Huh, I wouldn't have guessed that," Tan said.

"I'm not accusing anyone of anything yet," Sin said, "but I don't like Sec's behavior, and I'm worried about Cos. Do you think you and I can go to the game and be there just in case? It would be good if her friends were there."

Tan smiled compassionately. "Of course, I'll go. Thanks, Sin. I didn't think twice about there being a problem. I like basketball anyway."

"Great," Sin said. "I'll get the tickets, but I don't think we'll have the great seats that Csc and Cos will have."

CHAPTER 32

• • •

The stands in the stadium in the Q4 district were noisy with fans, students, and visitors purchasing their food and drinks and getting to their seats, ready for the basketball championship game.

Cos and Csc climbed over a few people already in their seats to get to their own seats.

"I'm so hungry," Cos said, digging into her nachos. "Do you want some?"

"No, thanks," Csc said. "I may be too full after eating this foot-long corn dog."

The game began with the team introductions, each player coming onto the court one at a time. As the announcer said, "At point guard, a senior, number four,

Secant," Sec jogged onto the court, and the crowd waved and cheered.

Csc looked at Cos to see her reaction to Sec, but Cos was paying attention only to her nachos.

"Sec likes you a lot," Csc said, raising her voice enough to be heard over the crowd.

"Well, Sec has no shortage of girls who like him," Cos said.

"Why don't you like him? Look at him. He's a star. I wish someone would like me the way he likes you," Csc said.

"It's hard to explain, I guess," Cos said.

The referee tipped off the ball, and the action began.

● ● ●

Even though they were much farther up in the stands than Cos and Csc were, Tan and Sin were getting into the game. While Sin was keeping one eye on the game and one eye in Cos and Csc's direction, Tan sighed each time a player missed a free throw shot. She couldn't understand why players forgot what was obvious to her.

If the basketball is shot so it can arc into the basket, it has a higher chance of going in. The ball sees the basket hoop as a circular shape, and it can fall straight in. If the ball is shot straight instead of in an arc, it will see the basket hoop as an elliptical shape with a shorter minor axis than the basketball's diameter and will be less likely to go into the basket and more likely to bounce off the back rim. A circular basketball hoop has greater area for the basketball to fall into than an elliptical basketball hoop. In general, a shot tossed at less than a thirty-two-degree angle of elevation will not follow an arc pathway.

"What are they doing? Why are the players making such lazy free throw shots?" Tan said in exasperation.

"Don't worry," Sin said. "Sec usually shoots his free throws at an angle of elevation of forty-five degrees."

"He understands," Tan said. "That's why he's developed his elite free throw percentage greater than ninety percent."

"I wish he had the same elite status in the skill of dealing with people and relationships as he has in basketball," Sin said. "I understand why Sec is interested in Cos. He sees she is good for him, but he doesn't understand. He is not good for her. He is not going to attract Cos. People attract those on their levels. Mediocrity attracts mediocrity. Oppressors attract those who enslave themselves. If you want to attract someone who rises above mediocrity, you must rise above mediocrity yourself."

"Do you still think Cos is in a vulnerable position here at the game tonight?" Tan asked.

"I hope not. I don't trust Sec, and I believe he's using Csc," Sin said.

"It wouldn't surprise me if she let herself be used by Sec," Tan said. "She has a lot of insecurities. She would do anything to be accepted by anyone. I always wondered why she felt she had to settle for mediocrity. She could be so much better than that. I never told her that, though. I'm inwardly too impatient with downers. I don't want to be dragged down with them, but I sometimes regret not encouraging her."

At halftime, Sin and Tan bought chili dogs and then spent the second half of the game discussing, cheering, and analyzing passes, plays, and shots from their own trigonometric viewpoints. Sin continued to glance in Cos and Csc's direction.

"How's your project coming along?" Sin asked.

"It's been a challenge. I'm trying not to get too overwhelmed over a mistake that was made," Tan said, knowing Sin would understand. The extent of logic and sense between them would ensure they got along well.

"Well, if you've only made one mistake with the size of the project that you're tackling, I'd say you're doing very well," Sin encouraged.

"Sometimes I wish we could work by ourselves on all our projects. When other people get involved, disappointing things happen," Tan said.

"Hmm. Sounds like you're a great candidate to run your own business," Sin said.

Tan looked at Sin. "I know. I like things to be accurate, and I don't like to waste time. I like working with other people if they are responsible, but not everyone is, and it drags things down. Do you experience that?"

"Yes, I know what you mean, Tan. Believe me; I do. I have similar tendencies, but the thing is, I want to be a leader, which means influencing for the better. It's good to be passionate about our work, our skills, and to accomplish good things. However, if my work is to mean something, I want those in my sphere of influence to be elevated in the process. There are also a lot of ideas to be gained from listening to others' thoughts, even if personalities don't always jive."

Tan pondered what Sin had just shared with her for a moment. Then, laughing, she said, "I agree in theory, Sin. In practice, how do you develop the patience?"

"Well, you're so good to Cos," Sin observed. "Why?"

"Cos doesn't exactly make it difficult to be around her. She's patient and responsible. But you know something? She doesn't try to control me or manage me. She doesn't bring out the rebel in me," Tan said. "I guess this is the first time I've tried to put into words my friendship with Cos."

"Interesting," Sin said. "Even though you are hard on other people, I think you have the same high expectations of yourself, so no one can say you don't have credibility."

"I'll clue you in on a little confession," Tan said with a wink. "My asymptotes give me a little insecurity. I actually do feel undefined at times. Maybe that's why failure and making mistakes are so hard to think about."

"Did the person who made the mistake on your project acknowledge it and apologize?" Sin asked.

"He knows he made the mistake, and yes, he did apologize, although he doesn't like to say those words," Tan said. "He assumes that flashing a charming smile will rebuild trust."

"I see," Sin said. "And I understand. It reminds me of the cliché, 'Just because I didn't say anything doesn't mean I didn't notice.'"

"Exactly! What makes you influential, Sin, is that you know how to deal with people, warts and all. It's hard to accept my own weaknesses, much less everyone else's. How do you—" Tan broke off as she looked at the court near the very end of the fourth quarter. "See what I mean?" Tan pointed to a player who had just missed a shot; the crowd was anxious, thanks to a close score. After several passes and some dribbling, Unit's team attempted a shot, but it bounced off the back rim. "That player could have made the basket if he had tossed the ball with an arc."

"The game is almost over, with the other team leading by only one point. This could get interesting," Sin said.

With ten seconds left on the clock, a couple of passes were made, and the ball landed in Sec's hands beyond the three-point arc. The crowd grew louder and slowly began to stand as Sec stood in place. He bounced the ball a couple of times and tossed it at the perfect angle for his height and distance from the basket.

With Tan's keen eye for seeing results from the use of a right triangle, she knew Sec had just won the championship game for the team and the university. As the ball spun in the air, making its way along a smooth, arced pathway, the clock ran down, and the buzzer sounded. It seemed the crowd in the stands held their breath and froze during the second it took for the ball to fall cleanly through the basket without even grazing the rim.

Heads turned to the scoreboard as Unit University's score increased by three points; the team had won the championship game by two points.

The stadium erupted, with fans cheering hysterically. Colorful streamers and confetti exploded into the air and fell gently to the floor.

Cos and Csc stood cheering while close family and friends of the players were invited onto the court in celebration.

Out of nowhere, it seemed, Sec appeared, holding out his hand to Cos to take her onto the court. Cos froze, not wanting to go.

"Just go," Csc said gently, pointing to the jumbotron showing that the camera was pointed directly at Cos and Sec. Cos summoned up the courage to follow Sec onto the court.

The players, with girlfriends beside them, were surrounded. Cos felt Sec's arms go around her and lift her up. She discretely avoided his attempt to kiss her and kept cheering, waiting for the moment when she could get away without drawing any attention or causing any embarrassment.

After several minutes, which felt like an eternity to Cos, the players and their girlfriends moved off the court. "Party time!" Cos heard someone shout. Sec took Cos's hand and pulled her along with him and the rest of the team.

"So you're Sec's new girlfriend?" Cos heard someone say walking next to her.

"Oh no, he's just being nice," Cos said to the girlfriend of another player.

"Sure. Lucky you! Plenty of girls would love to be in your position now," the girl said.

"Where are we going?" Cos asked.

"We're going to celebrate! Come on!"

● ● ●

Csc stood by herself, no longer cheering. Once she saw Cos go onto the court with Sec, she waited for people to move so she could leave. She then heard someone calling her name.

*The Trigonometry Tryst*

"Csc, wait!" Sin and Tan made their way down from the stands and finally caught up to Csc.

"Hi. I didn't expect to see you here," Csc said.

"We didn't have the seats you and Cos had, but it was a great game, wasn't it?" Tan said.

"Leave it to Sec to pull off the win," Csc said, her nervous giggle slipping out.

"Csc, do you know where Sec and Cos went?" Sin asked.

"No, I don't. Why?" Csc asked.

"We just want to make sure Cos is OK. I'm sure she planned to leave with you," Sin said.

"Maybe. But she went willingly. She didn't have to. Besides, who would turn down being invited onto the court personally by Sec after a win like that?" Csc asked.

"Are you sure you don't know where the team is going to celebrate?" Tan asked.

"Of course I'm sure," Csc said defensively.

"OK, we'll see you later, Csc. I'm glad you had such great seats," Tan said.

Sin and Tan turned to leave. Then Sin turned back around and said to Csc, "Do you want to come with us? You and I live in the same district. We'll make sure you get home OK."

Csc hesitated.

"Come with us," Tan encouraged.

"OK. Thanks," Csc said shyly.

CHAPTER 33

● ● ●

Sin, Tan, and Csc made their way out of the stadium at $7\pi/4$ and began walking counterclockwise along the circumference of the university.

"Do you two want to pick the bus up at the 0° gate?" Tan suggested. "I live in the Q1 district, so I'll just keep walking. Maybe Cos will be home by then. I keep checking my phone, but I haven't seen a message from her yet."

"Sorry," Csc said. "I'm asymptotic at the 0° gate, so I can't pick up the bus there. I don't mean to be any trouble."

"It's OK, Csc; we just forgot," Sin said. "We'll take—"

Just ahead at the fraternity house at $11\pi/6$, the sounds of partying could be heard. The streetlamps and outside lights of the house silhouetted the people out on the lawn. As the trio moved closer, they could see this was where the basketball team had come to celebrate.

"Cos is probably here," Sin said.

"Unless she got away from Sec already," Tan said. "Are you all right, Csc?" Tan thought Csc was looking anxious.

"Yes, I'm fine." Csc said.

Walking past the point of light where silhouettes faded and people's features came into view, they recognized team players and their guests. Beyond a couple of tall basketball players, Cos emerged. It looked as if she was walking away. A hand reached out and grabbed her arm.

"Let go of my arm," Cos said fiercely.

"How dare you leave after all I've done for you tonight," Sec hissed. He was obviously drunk.

"Leave her alone, Sec," Sin said in a serious tone.

Sec's head jerked in the direction of Sin, Tan, and Csc, his glassy eyes trying to process the presence of the unexpected guests. Looking at Sin, he snarled, "What are you doing here? Who invited you?" He then glared at Tan, and when he saw Csc, his face contorted into an even-angrier expression. "You brought them here! How dare you!" Sec slurred.

"I said let her go," Sin said.

"Now," Tan said.

Sec laughed at Tan. "And who do you think you are?"

"Cos's loyal friend. Something you're not," Tan said evenly.

Sec didn't like her piercing green eyes. They were penetrating and powerful. His grip on Cos relaxed, and Cos yanked her arm free. She moved quickly to stand by her friends.

"I'm tired of your always getting in the way," Sec said as he moved toward Sin and shoved him.

Remaining on his feet, Sin quietly said, "Let's go."

Cos, Tan, and Csc turned to leave.

"You're not going with them, are you?" Sec accused Csc. "You're staying with me."

Everyone turned to look at Csc. She could feel them watching her. Different scenarios played out in her mind. The security, peace, and friendship she felt with this group conflicted with the oppression she felt in the presence of Sec.

"Hey, guys, since I'm asymptotic at the 0° gate, I'll just head back this way. Thanks for asking me to come along," Csc said to Sin, Tan, and Cos with her back turned to Sec so he couldn't hear her.

Csc moved away from the group, unblocking Sec's line of sight to Sin. Unhinged at the sight of Sin and Cos standing together and the girl with eyes that made him look away, he lunged toward Sin again, this time planting a fist in Sin's face. Sin fell backward. Before Sec could strike Sin again, one of Sec's teammates jerked Sec back.

"What're you doing, man?" Sec's teammate asked.

"Stay out of this," Sec said, turning toward Sin, who was on his feet again, blood dripping from his lower lip.

*The Trigonometry Tryst*

"Knock it off, Sec. You're embarrassing the team."

Sec turned toward his teammate and laughed incredulously. "Embarrassing the team? Me? Who shot the three-pointer that won the game tonight? Me. Not you. It was me!"

The teammate went over to Sin, who was trying to convince Tan not to call the campus police. "I want to apologize for the team. I know you can fight your own battles, but if Sec behaves this way again, I will report him to the police myself, championship or no championship. This put a damper on our celebration tonight. Can I make sure all of you get home OK?"

"We'll be fine," Sin said, dabbing at his lip with a tissue from Cos's purse. "Thanks for sticking up for us and being an example for the team. Good luck."

"I can't tell you how relieved I am to see you," Cos said to Sin and Tan as they turned to continue walking counterclockwise toward the Q1 district. Csc had disappeared.

CHAPTER 34

● ● ●

"Meet me at the café for lunch," Sec texted.

Csc sat in class looking at the text on her phone, feeling an unexpected numbness. Normally, she would feel ecstatic about having lunch with Sec. *What's happening in me?*

Csc walked into the café, found Sec, and sat down at the table. He carelessly dropped the university newspaper in front of her.

"Check it out," Sec said. A photo of Sec shooting the winning three-point shot and a championship story filled the front page. Farther down, at the bottom of the front page, was a photo of Sec lifting Cos on the basketball court after the game was over.

Csc looked at Sec. She didn't understand his cocky grin. "Why are you smiling? You messed things up pretty well at the party."

Sec's grin turned into a grimace. "Well, I wouldn't have if you hadn't brought Sin and what's-her-green-eyed-face to the fraternity house. What were you thinking?"

"I didn't bring them. I followed them," Csc said, defending herself.

Sec eyed Csc for a few moments and then said, "I knew you couldn't betray me. We're too close for that to happen, aren't we?" He waited to see Csc's usual blush, but there was no reaction from her at all.

● ● ●

"I'm sure you saw this today," Tan said when she walked into the apartment at the end of the day. She set a copy of the university paper on the kitchen counter.

Cos was warming some leftover soup on the stove. "I saw it, and so did everyone else. You wouldn't believe the comments I got today. Everyone thinks I'm Sec's girlfriend now. I studied here today. I just wanted to come home."

● ● ●

"I'm posting my own Invitation to Meet," Sec said to Csc.

"Why do you have to do that?" Csc asked. "You'll just end up with a group of groping girls, the same ones who showed up to your last Invitation to Meet. Why not pick one of them? Or have you dated all of them already," Csc stated more than asked.

"I want to keep seeing Cos," Sec said.

"You're not seeing Cos. She doesn't want to go out with you," Csc said, rolling her eyes. "She's attracted to Sin's profile, and she obviously wanted to be with him instead of you at the party. What are you going to do, impersonate him again?"

"No. I'm going to set up an Invitation to Meet on the same date, time, and location of his Invitation to Meet," Sec said.

"You don't know the location," Csc said.

"No, I don't. But I will, because you will find out," Sec ordered.

"Sec, he hasn't posted it. I can't hack into his account again to look for something that isn't there," Csc protested.

"Figure it out from the clues. From that dumb telegraph story he thought he was so clever to post. Then I'll set my meeting up at the same location. Crash his party. Then Cos and he won't spend time together, will they? That's assuming Cos knows the location and will be there."

"And Cos is going to choose you over Sin at this crashed party?" Csc prodded.

"At least they won't be alone, and she'll get a chance to see me again. It will take the spotlight off Sin," Sec said.

"Sec. Come on. Just being in your presence puts an even-greater spotlight on Sin where Cos is concerned," Csc said, growing impatient.

"What!" Sec's face contorted into a familiar threatening countenance—which Csc realized she didn't find threatening anymore. She thought he looked silly, and was surprised by her lack of emotional response to accommodate him.

"She doesn't want to date you, Sec! Have you already forgotten what happened at the party, or were you too sloshed to remember any of it? You think you can keep Cos from looking at Sin. How are you going to manage that?" Csc asked.

"You mean, how are *you* going to manage that?" Sec countered.

Csc looked at him. "What now?"

*The Trigonometry Tryst*

"Besides, you owe me one," Sec said.

Csc raised her eyebrows questioningly.

"You didn't stay with me at the party," Sec accused.

"I didn't go with them either," Csc said.

"Of course you didn't, but I told you to stay with me," Sec said. "Your reputation would have improved if you had stayed."

CHAPTER 35

● ● ●

THE CLOCK TOWER CHIMED TEN times, telling the students where they should be at this moment.

Csc sat on a stone bench in view of the clock but didn't look away this time when she saw Cos coming toward her.

"Hi, Csc. You look like you were in deep thought," Cos said, smiling at her. "Hey, I'm sorry about leaving you behind at the game when Sec came and took me onto the court. I wish I had stayed with you. That was the best part of the game."

"It's OK. Is Sin OK?" Csc asked. "I didn't see him in the café today. He's probably lying low if he has a fat lip."

"I'm sure he'll be fine. So you saw all that," Cos said. "When we were leaving, we didn't see you anywhere. I'm glad you made it home safely."

A thought floated through Csc's mind. *I wonder if Cos has figured out the location of the meeting.* "I made it home just fine. Thanks. Crazy, wasn't it? I realize I'm changing the subject, but I was thinking about that mystery Invitation to Meet posted on Unit Match. The one that is missing a name and location," Csc said lightly.

"You were?" Cos looked surprised. "I've been obsessed with that Invitation to Meet since I saw it. Are you attracted to his profile too?"

"I just like the challenge, I suppose. To see if I can solve it," Csc said indifferently.

"Have you come up with anything? I mean, would you want to share information?" Cos asked.

"Sure, but don't think of me as any kind of competition. I don't even plan on going to the meeting. I just want to solve the mystery," Csc said.

"Well, so far, I'm thinking that the Invitation to Meet will be in the Q1 or Q2 districts, since his profile says he feels positive in those two districts," Cos said.

"That's logical," said Csc. "I've been wondering if there is any meaning in the date and time posted for the meeting. I mean, 8:07 p.m. is a geeky meeting time, isn't it?" Csc and Cos both laughed.

"Yes, it is!" Cos blurted out, still laughing. "But how can we get a clue from the time?"

"I've been staring at this clock for a while now. The time on a clock moves through a period every hour, right?" Csc mused while looking at the clock tower.

"Yes!" Cos jumped up. "So seven minutes after the hour would be forty-five degrees into its period. Well, seven minutes and thirty seconds, to be exact, but that would be too geeky."

"Exactly," Csc agreed. "'Meet me at 8:07 and thirty seconds p.m.' Ha!" Csc mocked. They laughed again.

Cos was motivated. "So let's look at the date to see if we can use the same logic."

"April thirtieth. That's during a weekend this year. Any connection there?" Csc considered.

"Hmm, I don't see one off the top of my head," Cos said, taking out her phone to use the calculator. "April thirtieth would be the one hundred twentieth day of the year, which would make it almost thirty-three percent into the year, or one hundred eighteen degrees into the earth's orbit around the sun, assuming that the period starts January first."

"That's clever," Csc said. "Still, I don't see a correlation between the date and time yet to point to one location. Do you?"

"Not yet, but I'm glad to be able to talk about it with you," Cos said, relieved. "Do you have my number? We can text each other when we come up with ideas."

After exchanging phone numbers, Cos looked at Csc and asked her, "So how are you doing? I didn't ask you at the game. Have you found a roommate yet?"

Csc looked down. "No, not yet, but I haven't been looking very hard either."

"Would you want to live in the Q1 district? I can let you know if anything opens up where Tan and I live," Cos said helpfully.

"I like the Q1 district, but how can you afford to live there?" Csc asked.

"I saved almost everything I earned with my jobs in high school. Plus, I was able to get a scholarship. That helps a lot. I haven't had to take out student loans yet," Cos explained without any arrogance. "Did you apply for a scholarship? I know you're really smart."

"I didn't even think about it," Csc said, shrugging. "My parents never advised me on anything. I just went off to college without knowing what to expect, really. I mean, I always knew I would go. I'll have student loans to pay back."

"You'll do well, though, so you'll be able to pay them back quickly, I think. Are there any openings for an aide in the computer science lab? You'd be good at that," Cos suggested.

"You think so? I've wanted to try. I just feel shy, I guess," Csc said.

"Well, I think you should go for it, and I will let you know when an opening comes up in our apartment building. I know we graduate in a few months, so you may not think it's worth it to move now, but you can decide when the opportunity comes up," Cos said.

"Thanks, Cos."

Cos went on her way, but Csc sat for a while longer, reflecting. *Cos feels safe. When she complimented me, I felt she was sincere. It doesn't feel like that with Sec. He compliments me because he wants me to do something for him.*

CHAPTER 36

● ● ●

Back in their apartment in the Q1 district, Cos told Tan, "I talked with Csc today."

"How's she doing?" Tan asked. "Did you ask her why she disappeared at the party? We would have made sure she got home OK."

"She didn't really want to talk about it. She's OK, I guess. There's something about her, Tan. She's very intelligent. If she can overcome whatever it is that's holding her back, she could do really well. I can't put my finger on it, though. I realize she appears awkward, but I think it's because she doesn't believe in herself. Do you think she's been put down all her life? She seems to behave like she's secondary to everyone else," Cos considered.

"I agree with you," Tan reflected. "That's kind of what I sensed when I was rooming with her. I regret I didn't have the patience to take the time to just sit and talk with her now and then. Her weakness bothered me. I wasn't mean to her, but I didn't give of myself to her either. I could have," Tan said honestly.

Cos listened, processing what Tan had just said. "Guess what. She's trying to solve the mystery location for the Invitation to Meet I'm working on."

"Seriously?" Tan remarked, her face brightening.

"Yes," Cos said. "She and I were discussing ideas. We're going to work together on it. I'm excited about working with her. I think I can encourage her too, if I can find out what makes her feel so miserable and nervous most of the time."

"So if you both figure it out, will that be awkward with the both of you showing up at the meeting? Or maybe it will be safer, just in case you don't end up liking the guy after all," Tan joked.

*The Trigonometry Tryst*

"Oh, she said she wasn't interested in the guy so much; she just wanted the challenge of solving the mystery of the secret location," Cos said.

"Huh, that's interesting," Tan said with a puzzled look.

CHAPTER 37

● ● ●

Cos pulled her jacket more tightly around her chin. The blustery weather put her in the mood for hot chocolate. *With some cinnamon in it.* She smiled to herself, hearing her shoes splash a puddle left by another rainstorm. Cos breathed in the fresh air that smelled of the cleansing rain and heard the faint, slow roll of thunder.

April 30 was getting closer. Four more days, but she still did not know the location of the meeting. Cos looked up at the sky as she walked, knowing that she wouldn't see any part of the moon illuminated. The new moon was under way, but it would have been hidden by the storm clouds anyway. She suddenly remembered that dreadful Invitation to Meet where Sec, the last person she wanted to see, was behind the meeting. She remembered the coincidence she had observed, relating the location at the 0° gate and the new moon insignia overhead to the actual phase of the new moon occurring that day. She hadn't thought about looking to see whether there was a correlation between April 30 and the phases of the moon.

Her mind stimulated, Cos stopped to look at the Naval Observatory's website on her phone to find that on April 30, the moon would be around 25 percent illuminated that afternoon. *I don't see a relation. Wait! When the moon is twenty-five percent illuminated, where will it be on its periodic path around the earth?*

Cos did a quick calculation on her phone. Assuming a perfect orbital circle for the calculation, when the moon was 25 percent illuminated, it would be 12.5 percent into its periodic path. And 12.5 percent equated to forty-five degrees from standard position, or the date of the new moon!

The forty-five-degree reference for the date matched the forty-five-degree reference for the time of the meeting, 8:07 p.m., the position of the minute hand on the clock into its new period. *Csc's idea. Could 45° or $\pi/4$ be the location on campus for the Invitation to Meet?*

Cos's heart was pounding. She called Csc to explain her idea.

● ● ●

Cos decided she would send the mystery person a message again to see if he would tell her whether she had determined the location correctly.

Hi,

I'm the one who sent you the message earlier because I thought you had forgotten to list the location for your Invitation to Meet and who you were. You sent me a note back with a challenge to figure it out. I've read "The Telegraph" many

*The Trigonometry Tryst*

times. I hope I've "heard" you. I would really like to meet you. I have a guess—tell me if it's right. My guess is the meeting place is $\pi/4$. Is that right? If it is, will you tell me your name now?

Cos

Cos was anxious to get a response. The meeting was just a few days away. When she got a response that evening, all it said was,

Cos,

Who am I? I am the same as you at the tryst.

Cos was both excited and annoyed to receive this response. There wasn't very much time until the meeting, and it was still a riddle. *Is $\pi/4$ the location or not? And what did he mean by "I am the same as you at the tryst"?* She decided to look at it again in the morning with a clear head.

When Cos woke up ten minutes before the alarm was to go off, she considered the new information. *"Who am I? I am the same as you at the tryst."*

*Well, he didn't deny the location. Is the location I guessed the tryst? Is he just answering my question of who he is now because I already guessed the location? We are the same at the tryst. Same what?* Cos pulled up his profile again and analyzed it for anything she hadn't noticed before.

Pulling up her own profile to compare side by side with the mystery person's profile, Cos was still at a loss.

*We are the same in almost every category except right-triangle definition and unit-circle definition. Wait. We are the same at the tryst. Does he mean I guessed right, and we are the same at $\pi/4$?* Cos decided to analyze their unit-circle definitions at $\pi/4$ to see what she would come up with.

Looking at their unit-circle definitions, Cos saw that hers was x/r and his was y/r. *So x and y are different. Could they be the same at the tryst? At $\pi/4$? On a unit circle,*

*The Trigonometry Tryst*

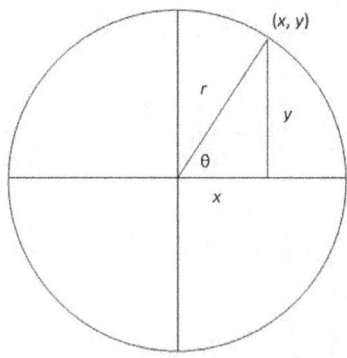

*r, or the radius, equals 1, so x equals y at π/4; x and y are the coordinates (x, y) of the address π/4. They are the same at π/4. Wait; x equals cosθ, and y equals sinθ. Cos π/4 equals √2/2. Sin π/4 equals √2/2. Sin. Sin! This is where we are the same. It's Sin!*

Cos sent Csc a text and invited her to come over to her and Tan's apartment to work on the problem.

$$\cos\theta = \frac{adj}{hyp} = \frac{x}{r}, \sin\theta = \frac{opp}{hyp} = \frac{y}{r}$$

On a unit circle, $r = 1$, so

$\cos\theta = x, \sin\theta = y,$

$(x, y) = (\cos\theta, \sin\theta),$

$\cos\dfrac{\pi}{4} = \dfrac{\sqrt{2}}{2}, \sin\dfrac{\pi}{4} = \dfrac{\sqrt{2}}{2}.$

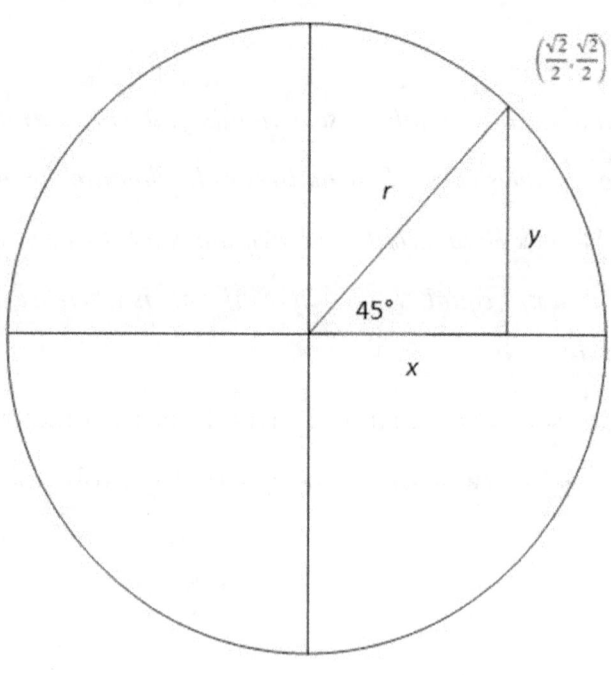

CHAPTER 38

● ● ●

When Csc knocked on the door, Cos opened it to see Csc smiling and holding a quart of intense chocolate ice cream.

"You are a lifesaver!" Tan said from across the room, going to get bowls and spoons. The three of them sat around the small table in the apartment.

"Tan, are you trying to figure out this mystery meeting too?" Csc asked.

"You mean the tryst? No. At first I thought this guy was loony. I thought he just forgot to include the missing information, which doesn't make a great first impression, but then I thought his strategy was clever when we discovered he intentionally made it a mystery. I like 'The

Telegraph.' Cos is attracted to his profile. He's all hers. I just hope he's not a disappointment. That is, if you guys get to meet him. Any luck figuring it out?"

Cos looked at Tan and Csc, trying to hold back her excitement. "It's Sin!" she blurted out. "At least I'm pretty positive it is."

"What? You're just telling me this now? What makes you think so?" Tan asked, slurping a spoonful of ice cream.

"Well, Csc figured out the first part. The meeting time is 8:07 p.m. A strange choice for a meeting time. It turns out that the meeting time is a clue to the meeting location. Seven minutes—well, seven minutes and thirty seconds to be exact—is forty-five degrees into the period of an hour," Cos explained.

"I'm following," Tan said.

"Csc and I thought that if there is a clue in the meeting time, there is likely a clue in the date as well," Cos said.

*The Trigonometry Tryst*

Csc and Tan both nodded, ice cream melting on their tongues.

"Csc and I tried to see if the date was also forty-five degrees into the earth's period around the sun or the calendar year. We couldn't make it work, so we didn't know if $45°$ or $\pi/4$ was the right location or not.

"I was walking and looking up at the moon and realized that on the date of the meeting, the moon would be twenty-five percent illuminated, or forty-five degrees into its monthly orbit around the earth, which completes a periodic cycle each month. So $45°$ or $\pi/4$ is the common clue between the meeting date and time!" Cos said.

"No argument from me. I just hope two clues are enough," Csc said.

Tan high-fived her friends. "Now, how is this linked to Sin?"

"This next part is what I wanted to wait to tell you until Csc got here to see what you two think," Cos said.

"When I e-mailed the mystery man to ask if the location was correct, he didn't confirm or deny it. He just replied with another riddle that said:

'Who am I? I am the same as you at the tryst.'"

Csc and Tan both looked at Cos blankly.

"I'm beginning to think the riddle is a confirmation that the location you guessed is correct," Tan said.

"Me too," Csc said. "The location has been figured out. Now it's time to figure out his identity. I think the riddle now would read,

'Who am I? I am the same as you at $\pi/4$.'"

"That's what I think too!" Cos could hardly contain her excitement. "If we are the same at $\pi/4$, then our unit-circle definitions and our right-triangle definitions should each have the same result at $\pi/4$. Do you agree?" Cos asked.

Csc and Tan nodded in unison.

Cos grabbed a pencil and a spiral notebook to derive the results of the sine and cosine functions at $\pi/4$ using the 45-45-90 special right triangle:

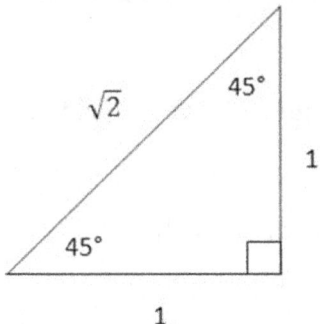

"No argument. I would say it is Sin too. Nice work, both of you," Tan said.

Csc kept her knowledge of Sin's identity to herself, as she was not proud of how she had arrived at this knowledge. "Your persistence paid off, Cos."

"We all did it together. That's what happens when we share ideas." Cos beamed. "I can't wait until the meeting. Are you two sure you don't want to come?"

"And intrude on what promises to be the most romantic night of your life so far?" Tan winked.

"Csc, you are deep in thought again," Cos observed. "Are you having second thoughts about how we solved this mystery?"

"No, I was just wondering what my complement would be like. My cofunction." Csc blushed.

Cos's eyes widened with the idea. "Let's see! Then we'll see what yours looks like, Tan."

Cos turned to a fresh sheet of paper while Tan dished up more ice cream for everyone.

"You have good taste, Csc. I love this ice cream," Tan said.

"We'll have to do this more often," Csc said with confidence.

"What is your right-triangle definition, Csc?" Cos asked.

"Hypotenuse divided by opposite side," Csc said.

"Hmm. Interesting. That's the reciprocal of Sin's," Cos reflected.

"I know; I've always felt like his reciprocal, secondary to him," Csc said.

"OK, at $\pi/4$, you are $\sqrt{2}$," Cos said, looking up from the work she had just done on the paper tablet.

"Since ninety degrees minus forty-five degrees equals forty-five degrees, my complement or cofunction should also be $\sqrt{2}$ at $\pi/4$," Csc reasoned. "I wonder how to find out who is $\sqrt{2}$ at $\pi/4$."

"Well, if you need a computer to do it, you would be the best one to figure it out," Tan said. "Besides, I've spent very little time on Unit Match to know all the options it provides."

"Thanks, Tan. Actually, I do have an idea. I think that you can do a query on Unit Match. I'll search for those students who are $\sqrt{2}$ at $\pi/4$. Does anyone have a computer or tablet I can borrow?" Csc asked.

"Here, use my laptop. It's already booted up," Tan said, bringing the laptop over to Csc. "I'm starting to get anxious about finding out who my complement is now. Imagine that!"

Cos tried not to laugh with her mouth full of ice cream.

A moment later, Csc looked up, and both Tan and Cos watched her with concern.

"Csc, your face just turned white," Tan said.

"I…I guess I didn't expect the results that came from the query. There's only one result," Csc said.

"Who is it? Anyone we know?" Cos asked.

"Sec," Csc said, leaning back in her chair.

"Oh," Cos said.

"Oh," Tan said.

There was silence at the table for a minute until Tan started to laugh. Her laugh built up from a quiet giggle to a loud belly laugh, which infected Cos and Csc.

"OK, Tan, let's see how funny your complement is," Csc said when they all calmed down.

"I can't wait!" Tan said. "While you were doing the query, I determined that I am 1 at $\pi/4$, so query anyone who is 1 at $\pi/4$."

"Hmm. There is only one result for you too," Csc said.

"Who?" Tan and Cos both said in unison.

"Cot."

Tan and Cos looked at each other knowingly, and the roar of laughter ensued again.

Csc laughed with them, assuming Cot was as much a disappointment for Tan as Sec was for her. "I take it that Cot is not the one you hoped for, Tan?"

"No! Not in the least! Unit Match's algorithms are crazy. I was right not to spend my precious time on that website. This is a joke! Although I do have to admit the results worked in Cos and Sin's favor," Tan said, still laughing.

"Well, if you can forgive Cot, you may see some good points about him," Cos said.

"What happened between you and Cot, Tan?" Csc asked.

"Cos is right. What I just said about Cot and his being a huge disappointment isn't entirely true. What I said wasn't fair. There's a side to him I actually look forward to being around," Tan said.

After Tan summarized some of her troubled feelings about Cot for Csc, she shared with her friends something about herself she was becoming aware of.

"I know he has good points, but I've been doing some soul-searching, and I've had this problem all my life. When someone has crossed the line with morals, ethics, or bad manners, I feel like I have to show them that they didn't fool me. I want them to know they aren't going to wipe their boots on me. People think if you don't stand up for yourself or say anything, you have no problem with them or what they did. I don't know how to pretend

I like someone who has broken trust. I also don't know how to deal with warts and all. I don't know how to deal with people's weaknesses, even though I have plenty of weaknesses myself. I don't want to be dragged down."

Tan continued. "At the same time, if someone wants to work something out or is sincerely asking for forgiveness, I can move forward without looking back. When I feel I owe someone an apology, I can't wait to give it. I want it to be worked out and for us to move forward. It's crazy when people assume that letting time pass is the same as an apology. It's hard to move forward and trust them without ever having talked it through or worked it out. Things will continue to quietly fester. The relationship will be nothing more than an act going forward, with genuine feelings having dissipated. The offending person will feel he or she can treat you badly again.

"Look, I know I'm blunt and straightforward, but most people trust me for that. I just can't, like everyone else, make offending people feel good about themselves just to be polite. So I'll have some friends, and I'll have some enemies, I guess. I've got to be true to me. I don't think I've ever shared these feelings aloud before. To be fair to Cot, he did say he was sorry, and I feel the apology was sincere. He did try to make it right by helping me correct and resubmit my project. That's why I don't despise him anymore," Tan said.

● ● ●

Tan had trouble falling asleep. Her complement, her cofunction, was Cot. Yet he was her reciprocal. Their personalities and profiles intertwined so much, but she couldn't tell whether that was good or bad. He could drive her crazy with his "whatever" attitude. She liked to make plans and set goals. Yet she was becoming aware that more and more she was looking forward to seeing him again. It wasn't logical. Her heart was talking this time. *I just realized he is patient with my impatience. He accepts my warts.*

CHAPTER 39

● ● ●

SEC LOOKED AT CSC'S FACE. She was smiling in deep contemplation, and she didn't look troubled. He didn't know whether he liked that. He could sense something different about Csc that made him less confident he could manipulate her.

"You know where the location for Sin's meeting is, don't you?" Sec asked, certain of the answer.

Startled, Csc looked up at Sec and saw the wariness in him. Sec, in turn, noticed the calm in her eyes. He stifled a spark of respect and a hint of attraction to Csc. He was more comfortable with their relationship as it was already.

"I did," Csc spoke slowly. "Would you believe it's at the stadium? Address $7\pi/4$. A place you like to be anyway."

"How did you figure it out?" Sec asked, sitting down at their table in the café.

"The clues were in the time and date," Csc explained. "The time to meet is 8:07 p.m. Strange time, isn't it? It's because 8:07 is forty-five degrees into the period of an hour on a clock. Do you remember when you posed as Sin at the meeting at the $0°$ gate at noon on the day of the new moon in February?"

"Of course I do. You were brilliant to relate all the meeting details with Cos's passion for astronomy. I thought she would appreciate a time, a date, and a place that were all representative of the beginning of a new period. A new hour, the date of a new moon, and standard position on our campus. She probably didn't even notice," Sec said sarcastically.

*The Trigonometry Tryst*

"Well," Csc said. "Sin thought of the same concept. He also related his meeting details with an astronomical behavior."

"So how do the date and time work together to point to the stadium as the location of the meeting? You just said that the time of the meeting suggested the location would be at $\pi/4$," Sec said.

"No, it's more complex than that," Csc said, taking her pencil and drawing a diagram of the moon's illumination as it would appear on the date of the meeting:

"On the date of the meeting, April thirtieth, the moon will be twenty-five percent illuminated, which is twelve point five percent into the period around the circumference of the campus, which would be forty-five degrees from standard position or the 0° gate. But look. On the diagram, the moon is twenty-five percent illuminated on this date, but it is a waning crescent moon, which means it's twenty-five percent illuminated on the tail end of its cycle of its orbit around the earth. So instead of moving counterclockwise from the 0° gate, we should move forty-five degrees clockwise from standard position on the campus. This puts us at 315° around the campus from standard position, the 0° gate, and lands us at $7\pi/4$, the address of the stadium!" Csc said.

"Clever. Modify Sin's Invitation to Meet again and add the location you figured out. That will show him. Make sure to set up my Invitation to Meet with his same time, date, and location," Sec instructed.

*The Trigonometry Tryst*

Csc opened her laptop, marveling at the new sensation she was beginning to feel: being centered, anchored to something good. There was an Invitation to Meet to set up, and she'd better get to it.

**CHAPTER 40**

● ● ●

TODAY WAS THE DAY TAN would find out whose design had been selected for the campus building. The senior banquet for the architecture department was this evening, the same day as the tryst.

Tan hoped that this time, the meeting would be all Cos hoped for and she and Cos would both win tonight, each with her individual desires. She sat in the library analyzing the hexagonal windows above her. The brilliant sunlight exposed the two opposite triangles centered over each other, forming the six-pointed star.

Daydreaming, Tan pictured herself at the senior banquet. Dessert plates just being cleared, she pictured the department chair walking toward the podium. She and her classmates waited in anticipation as awards and recognitions were handed out. The most important award to her was having her design selected for the campus building. In her dream, the obstacles she had overcome didn't hurt her chances of winning. She heard the department chair announcing her name and calling her up to the podium as the room erupted in cheers and applause.

Tan knew she had done her best on the project even if she didn't win. She tried to prepare herself psychologically just in case. There was a good chance of that too, she felt, since she had had to recall her design after she had submitted it.

One of Tan's favorite quotes from Henry David Thoreau was, "Our truest life is when we are in our dreams awake." *Oh, how I want to be awake in this dream and have it come true.*

Tan noticed that Cot had entered the library. He was studying at a nearby table. She watched him with mixed feelings. She had become better friends with him as they worked on her project, even though they were reciprocals of each other. She enjoyed being around him, and in many ways, they complemented each other. She tried not to dwell on the thought that he could be the reason she didn't win tonight.

Cot looked over at Tan. He saw her looking at him, and he smiled softly. He came over to her table.

"Hi," he said.

"Hi, Cot."

"I saw you when I came in, but I decided not to bother you," he said.

"That's OK. I'm getting ready to leave anyway. We have our senior banquet tonight. I'm going to go home to get ready," Tan said.

"Really?" Cot paused for a moment. Then he asked, "Do you need to bring a date?"

Tan looked surprised. "No. I mean, spouses and dates are invited, but it's not required, of course."

"Of course not," Cot said with some awkwardness. "Are you bringing a date?"

"No, I'm not dating anyone right now," Tan said.

"Well, what if I were your date tonight?"

"I don't know. I mean, do you really want to be there?" Tan said, stalling.

"I'd like to be there with you. You've worked hard," Cot said.

"Well, if you want to. It starts at seven o'clock."

"I'll pick you up at six thirty," Cot said.

"OK," Tan said, smiling inside.

They packed up their books and went to get ready.

● ● ●

Tan and Cot took seats at a round banquet table next to several of Tan's classmates and professors and their spouses and dates. Cot was introduced to everyone at the table.

At the table was the professor Tan had retrieved her project from when she realized the survey mistake. She saw him look at Cot and then look at her. *Was it a mistake to bring Cot here? What is my professor thinking about?*

Tan chided herself for wanting to win so much she was embarrassed to be seen with someone who could hurt her chances. *Cos thinks I'm solid as a rock. If she could hear my insecure thoughts now, what would she think?*

Dinner progressed smoothly with polite conversation and people getting to know one another's spouses and dates. As dessert plates were cleared, the department chair walked toward the podium. Tan had played this scene out in her imagination so many times this year.

Applause greeted the awards handed out for freshmen, sophomores, juniors, seniors, and graduate students. Several professors' contributions were announced.

"And now, the moment the seniors have been waiting for," the department chair said into the microphone. "Each senior had the opportunity to submit a design for the campus building to be built at $5\pi/4$. They were given the purpose of the building, a budget, and some criteria. It was then up to the students to use what they have learned in the architecture program to create what they could envision. After studying each design, the committee selected a design that used the budget wisely, was accurate in the fundamental principles of architecture, used correct calculations, and enhanced the campus environment."

Tan's palms were sweaty, and her heart pounded so loudly she was surprised others couldn't hear it. She looked at her professor. He looked at her knowingly and winked.

The department chair continued. "The proposed plan that best complied with the criteria given was designed by Tangent. Tangent, would you come to the podium, please?"

The room erupted in applause. Tan's name echoed in her ears.

"Congratulations!" Cot said to her, clapping enthusiastically while Tan stood and walked to the podium.

A three-dimensional simulation of her design was projected on a large screen beside the podium. She hadn't seen the design of her work on this large a scale yet. *I'm in my dreams awake.* The department chair described some of Tan's accomplishments at the university and in the architecture program and then gave her a few minutes to talk through her design and the thoughts underlying her decisions to create the final building. As Tan summarized the architectural principles she had used in her design, she discussed some lessons she'd learned from an academic perspective. What she didn't acknowledge aloud was the growing realization that one lesson trumped all others: what she had learned about relationships by working with Cot.

As Cot listened to Tan describing her design, he detected her exhilaration and passion for her work. He reflected on their relationship so far and felt regret over his part in holding her back from accomplishing what she wanted. *I don't know how to be as structured or goal oriented as she is, but I want to help her, and I do want my word to mean something. I understand that now. I think I'm beginning to understand what it means to have substance. Will she ever know the influence she's had on me? Will she ever let her disappointment in me go? Can she forgive me, or will she always hold a grudge against me?*

Cot saw Tan looking at him as she spoke; her eyes held no accusation. As the evening wound down, guests began to leave, and the banquet room thinned out. Tan felt she had taken the opportunity to shake every person's hand. *I won. I won! I'm in my dream awake!*

Her professor stood at a table by himself, and Tan realized he had not congratulated her yet. He signaled for Tan and Cot to come over to the table.

"Take a seat, both of you, for a couple of minutes. I want to share something with you," the professor said.

Tan and Cot sat down at the table and waited eagerly for what the professor had to say. *Is he glad I won? Or does he think someone else deserved to win?*

"Tan, let me congratulate you on a job well done and well deserved," her professor said.

Tan couldn't help grinning from ear to ear.

"I know it wasn't your best day when you had to recall your design after submitting it; however, I want you to realize that this showed your competence as an architect."

Tan looked at Cot and back at her professor, taking in what he was saying.

"You recognized the incorrect survey reading. You looked at the slope of the plot and discerned that it was steeper than the survey reported. You took the initiative to measure the slope and anticipated the changes you would need to make in the design based on the corrected slope calculation. You never gave up."

The professor noticed Cot shifting in his chair uncomfortably.

"The ability to understand what you are looking at and to be able to discern the reasonableness of calculated results is a lesson that most professionals don't learn as early in their careers as you have, Tan. You were embarrassed, but you were actually proving your skill set. I'm proud of you," the professor said.

"Well, I guess the mistake I made in the survey was a good thing after all," Cot said jokingly. "I was the one who performed the survey and made the mistake, Professor. It wasn't Tan's fault."

Tan shot Cot a look. *Wow, he really has changed his tune.*

"Now, here's the second concept I want to talk with both of you about," the professor said. "Neither one of you will work alone in your careers. Whether you work as an employee or own your own business, you will always need to work with other people. Working with other people means accepting that people will make mistakes and problems will need to be worked through.

"Both of you will need to understand what your end goal is in every situation and the results you want to obtain. You will need to handle problems that come up accordingly. Relationships are important. Your technical skill sets will take you only so far. Your ability to work with people will take you as far as you want to go."

Tan and Cot sat thoughtfully for a moment.

Tan said, "Professor, what you say makes sense. It just seems that when some people say you have to work with other people, they mean you have to give in to everyone else's opinions, ideas, or even unacceptable behaviors. What if I really believe in my own idea or strategy and no one else has come up with ideas that seem better than mine?"

The professor smiled. "You've been thinking about this concept already. Every situation will be different, Tan. You will need to recognize the strategy you will use to accomplish your goals and keep relationships intact.

"You just mentioned an approach you will almost always use: asking other people for their opinions and ideas. As analytical as you are, you still won't always have the answers. Someone else may have a better idea or more experience. Whether an idea is your own or someone else's, the important things are that you ask and that you recognize the best ideas that will drive the results you want. If your idea is the best, stand your ground. You will know from experience if you need to compromise. Don't, however, stand your ground only to protect your pride and ego."

"I understand, Professor. Thank you for taking the time to talk with us. It helped a lot," Tan said.

Cot chimed in. "Since I've been working with Tan, I've been thinking about this too. Working with other people and keeping relationships intact doesn't mean there should be a lack of standards and expectations. If everyone did only what they felt like doing, nothing would get accomplished. A team or group with a purpose needs to decide what their goals are and agree upon standards. If someone decides they are not going to do what they committed to or behaves in a way that hurts the team, it's time to let that person go."

"Well done, Cot. People will help you, and people will fail you, just as you will help or fail other people," the professor said.

Cot looked at Tan sheepishly.

"Well, I think you both have learned from each other, haven't you? You two complement each other well," the professor observed and stood up.

"You both have the potential to be great leaders in your fields and within your spheres of influence. Remember what we discussed. Congratulations again, Tan. Your win was well deserved," the professor said. He shook Tan's and Cot's hands and left the room.

Tan and Cot looked at each other knowingly. As they left the banquet, Tan allowed Cot to take her hand.

CHAPTER 41

● ● ●

Sec was standing in front of the stadium at $7\pi/4$ right on time—at exactly 8:07 p.m. on the evening of the tryst. Hysterical girls started jumping up and down and running toward him. He smiled at them, expecting and enjoying the attention; however, while he was charming the girls, his awareness reached beyond them. Csc hadn't arrived yet, and Sin was nowhere to be seen. He couldn't wait to see the look on Sin's face. Cos wasn't there, so she probably hadn't figured out the location. *Csc is smart*. After ten minutes of indulging the girls with compliments, he became nervous. "There's going to be another meeting here tonight, girls," Sec teased them. "I can't let Sin steal you away, can I?"

"Sin?" one of the girls said. "We passed him on the way here. He must be having his meeting over at that alcove at $\pi/4$. He and Csc were sitting together! Csc!" The girls burst out laughing. "Crazy-eyed Csc together with Sin. What a match!"

Sec felt the heat move to his head, and he wasn't sure how long he could keep up the deception. He turned around to take a deep breath. Feeling the cool breeze on his face, he raised his head and closed his eyes.

When he opened his eyes, the first thing he saw was the moon. It was 25 percent illuminated, just as Csc had said it would be tonight. The diagram that Csc had drawn of the partially illuminated moon floated through his mind.

It didn't register at first, but suddenly something clicked. Csc's diagram had shown the illumination on the left side, a waning moon. As he stared up into the sky, he saw that the moon clearly was illuminated on the right side. *It's a waxing moon tonight, which means that twenty-five percent illumination is forty-five degrees into the moon's period around the earth, matching with the meeting time. The time 8:07 p.m. is forty-five degrees into the period of an hour around the clock. Sin and Csc are sitting in the alcove at 45° or π/4.*

"Look at that crescent moon," Sec said as he turned back around to the girls. "One of you look up the Naval Observatory on your phone. Let's see if they show a waning or waxing moon tonight."

"What?" one girl asked. "You want to look up data about the moon, now? Are you all right?"

Another girl pulled up the Naval Observatory data on her phone, which she handed to Sec. It was a chart of the percentage illumination of the moon by day for each month of the year. This day, April 30, the moon would be 25 percent illuminated. But it would be waxing, not waning. Sec could barely hear the noisy groupies as he looked at the information on the phone.

"Come on," one of the girls said, tugging on Sec's sleeve. "Let's get this party started!"

● ● ●

Sin was early. He sat on a stone bench at $\pi/4$ and felt the light breeze on his face. The moon and the streetlamp nearby provided all the light he needed. The stone bench was one of three stone benches equidistant from each other, the curve in each bench lining the inside circumference of a circular cove that terminated a radial sidewalk, which originated from the clock tower at the center of campus. It was a perfect outside conversation area. *Maybe Cos will show up. Maybe no one will. Cos was interested enough in the profile and "The Telegraph" to figure out the meeting location. Did she figure out who I was?*

He shifted forward to lean his elbows on his knees. *It's so quiet except for the cool breeze through the trees, the rumbling distant thunder, and the thoughts in my head.*

## The Trigonometry Tryst

Pulling his jacket collar up around his neck to keep from getting too chilled, Sin noticed that even with the sounds of the beginnings of a spring storm, the sky wasn't cloudy enough to hide the silvery light of the crescent moon tonight. He thought about the perfect location he had chosen. Everyone loved the Q1 district. It only made sense to choose this district. Students wouldn't be deterred by meeting in a district that felt negative to them. As he hoped with anticipation, he also became lost in his thoughts. Then he heard someone's shoe scrape the sidewalk.

The sound came from the direction of the streetlamp a few feet away. It was still early. *Someone may just be passing by.* He saw someone walking toward him but could see only a silhouette, as the person was on the other side of the streetlamp. He could tell by the hair that the person was a girl.

She came into view, and Sin wasn't sure what to think when he saw her. Csc was obviously walking directly toward him. *Is she here for the meeting?*

"May I sit down?" Csc asked as she deliberately walked right up to Sin.

*She's not just passing by. She wants something. Has she been crying? Her eyes never look relaxed.*

"Of course you can," Sin said compassionately as he moved over.

Csc sat down and huddled in her jacket. It was getting chilly. Soft, long blond curls went in every direction—a look that was unique to Csc and, Sin felt, could be endearing.

"What brings you over here tonight?" Sin asked patiently but with curiosity. "Are you responding to the Invitation to Meet here, or did you know there was one?"

"I know you posted an Invitation to Meet here tonight. I was hoping you would be early so I could talk with you privately," Csc said, looking at him quickly and then turning her head toward the ground with a sniff.

*She knew beforehand that it was me who posted the Invitation. Is she interested in me?*

"It was me," Csc announced quietly but directly, still looking intently at the ground.

A moment passed. "Go on," Sin said quietly.

"I was the one who hacked your Unit Match account and changed your Invitation to Meet details when Sec posed in your place with your profile description in February," Csc said stonily.

"I know it was you," Sin said, but not accusingly.

Startled, Csc looked at Sin.

"Why did you do it?" Sin asked.

"I'm not sure if you can understand." Csc turned her head away again to look at the ground. "How can you know what it's like not to belong or have friends?"

"You think I belong, huh? What makes you think that?" Sin asked. He was going to enjoy this conversation.

"You're good looking, confident, smart, respected on campus. People like you," Csc said. "I always feel so secondary to you."

Sin took a deep breath. "You know, Csc, I realize we are both still young, but one of the greatest misperceptions I've observed throughout my young life so far is that achieving success is easy. We don't realize the persistence, the dark moments people go through to try to be something better than they are. These moments can be painful but are worth it to transform them into better and more successful people than they were before.

"No one is left untouched. We all have our unique struggles. What makes the difference is how you react to and manage yours, mastering your emotions, channeling them to transform yourself, affecting relationships and your sphere of influence for the better."

Sin and Csc sat together in silence for a while. Leaves rustled on the trees. Csc's mind was turning. She was beginning to understand.

"So," Sin said, "you hacked into my account to belong? There must have been other people involved. You've been doing Sec's dirty work, haven't you?"

"Yes, but you knew that already," Csc said.

"Doesn't it feel better to get it off your chest, though?"

"Yes. Yes, it does," Csc said slowly as she looked at Sin. She started to smile, and with that smile came the beginning of confidence. Honest eyes looked at him and held his gaze for longer than a second.

"You're ready to take a different course, Csc. A course of your own choosing that will make you successful in life. You want something different. I can see it in your face. Sec belongs. Sec is good looking, a star athlete who may have temporary success, but he won't have lasting success. You sense that. That's why you're talking with me and why you owned up to what you did and made it right. I admire that in you," Sin shared.

The heaviness in Csc's heart began to slip away, and the relief was enough to make her want to cry. She saved it for later when she could be alone. She felt Sin's words. *I do want to take a different course. I can choose.*

"A question we are asked a lot at the university and will be asked especially when we start interviewing for jobs is, how do you see yourself in ten years?" Sin reflected. "Try it, Csc. If you haven't already, picture yourself in ten years. What kind of person do you want to be? What will you have become? I believe that when making choices about what to think about and what to do, you need to make sure you think about and do those things that will make that vision of yourself in ten years a reality."

"Thank you, Sin. Thank you for treating me this way after what I did to you, and for helping me." Csc could barely get the words out; she was so overcome with emotion. The feeling of hope and determination tasted so good. "I realize now I was willing to settle for mediocrity in my attempts to fit in. I have a purpose. I'm choosing greatness over mediocrity going forward."

"That's what friends are for." Sin reached out and hugged her. "I'm looking forward to watching you, Csc. With your intelligence and great hair, it won't take you long to rise fast."

Csc burst out laughing. It wasn't her nervous giggle. It was a beautiful, genuine laugh that came up from her innermost self.

Sin noticed how attractive she had become in the last little while he had just spent with her. *Inner attitude really does determine how attractive one is to others.*

● ● ●

Cos watched the silhouette of Sin and Csc hugging against the bright backdrop of her beloved moon. It was 8:07 p.m., and she was standing at the location of $\pi/4$, the long-sought-after location of the tryst.

## CHAPTER 42

● ● ●

CONFLICTING FEELINGS HAD COS AT a standstill watching Sin and Csc and processing what she was seeing. The relief she felt to find that the mystery person was Sin made her smile inwardly. There was a connection now with her attraction to the profile and her equally strong attraction to Sin. Knowing the two were one gave her intuition credit. How she had hoped!

Now that hope was tempered, though, in a more poignant direction. *Sin and Csc? They are so dissimilar.* Jealousy wasn't one of Cos's recurring weaknesses, but she felt it intensifying in her now, and she didn't like it. *I'm better than this! Besides, everyone is invited. More people might come. This wasn't an exclusive Invitation to Meet. But*

*Csc said she wouldn't be here! She wasn't interested in who the person was. Why did she deceive me?*

To keep her thoughts from spiraling, Cos moved into the light of the streetlamp. She was determined to behave well.

"Hey, guys," Cos said with as much cheer as she could muster without betraying the fragility she felt inside.

Sin and Csc looked up to see Cos enter the circular cove of stone benches, her teeth slightly chattering. "Hi, Cos!" Csc greeted her genuinely.

Forgetting about Sin for a moment, Cos was taken aback by a transformation in Csc's appearance. It wasn't a new haircut, clothes, or makeup. *What is it?*

"Csc, you look wonderful," Cos said warmly.

"Thanks. I feel great," Csc said without a nervous giggle to go along with it.

Cos's eyes searched Csc's for a moment, and Csc didn't look away. Cos felt power coming from Csc for the first time. *Csc looks beautiful and so intelligent. So this is what transformation looks like when someone chooses to have her strengths govern her weaknesses. What triggered this change? I want to be around her now.*

"So, Sin, Csc was my partner in crime. We figured out the location of this meeting together. I couldn't have done it without her. Her ideas inspired my ideas," Cos said.

"I wanted to figure out the location of the meeting, but for different reasons than hoping to meet my complement," Csc said. "Sin is my friend, and I feel you are my friend too, Cos." Csc stood up. "I'm going to give you two some space. Besides, it's getting cold! Thank you both for everything."

The trio stood up, and before they could say anything else to one another, they all turned in the direction of the sound of someone running toward them.

Sec burst into the cove, erupting in anger so unbridled that Cos stepped back instinctively. Sin took her hand and held it reassuringly.

"You betrayed me!" Sec roared at Csc with a crazed look in his eyes.

Csc didn't flinch.

"Sit down, Sec. Cool off," Sin commanded in a low, steady voice.

"You stay out of this, Sin!" Sec shouted, pointing a finger in his face. "You're nothing! You've got these two fooled."

Sec shot a look at Cos and then turned to Csc. "You screwed up! I'm dropping you, and you can go back to being the loser you are and always will be!"

Silence ensued, but not because anyone recoiled. The breeze picked up, and the thunder was close and louder now. Sec panted angrily, but his anger was met with silence, which turned to awareness.

Sec looked from Sin's steady brown eyes, to Cos's fixed blue eyes, to Csc's unyielding, confident hazel eyes, which unsettled him the most. A corner of Csc's mouth had even turned up without her breaking her resolute gaze on him. The group stood there until Sec's breathing calmed.

"Let's all sit down," Sin repeated calmly.

Sin and Cos sat down together on one of the three stone benches. Csc took a seat on the bench to their right. Not sure whether Sec was going to run or have the courage to stay, Sin, Cos, and Csc patiently watched him make the decision. A moment passed, and then Sec sat down—next to Csc.

CHAPTER 43

• • •

"Cot, would you mind coming with me? I want to stop by the cove at $\pi/4$ to see if Cos is still there," Tan said after they left the awards banquet.

"Sure, though it's a chilly evening to spend in the conversation cove," Cot said. "And it looks like it's going to storm."

"I know, but there was a meeting there tonight," Tan explained. "An Invitation to Meet was posted on Unit Match. Cos was interested in it, so I want to see if she's still there and make sure she's OK. If she's not there, I'll just meet her at home."

"Who is she interested in? Anyone we know?" Cot asked.

"I'm pretty sure it's Sin. I hope it is, anyway. She's been interested in him for a while," Tan said.

"You mean Cos didn't know for sure who she was meeting?" Cot was perplexed.

"It's a long story. I'll tell you on the way over there," Tan said.

● ● ●

Approaching the conversation cove, Tan and Cot could see four people sitting on the stone benches softly illuminated by the streetlamp. The lamp's warm orange-yellow glow enveloped the cove and everyone in it. Blackness surrounded the glow. Even the moon was behind the clouds now. As they drew near, Tan thought the scene looked like a campfire in the mountains on a dark night.

"Four people. That means Cos has some competition," Tan said teasingly. When they got close enough to make out the faces, Tan's teasing mood turned into one of concern.

"What is Csc doing here? She said she wasn't coming. She wasn't interested. And what is Sec doing here?" Tan wondered softly.

Cot was confused but remained patient.

"Tan! Cot! Come in," Cos said as the couple emerged into the glow of the cove. Tan gave Cos a questioning look. Cos's smile communicated that everything was OK.

"For an Invitation to Meet whose location had to be figured out, there are a lot of people here. I'm impressed. I should have made it more challenging." Sin grinned.

"Cos and Csc are the ones who figured it out," Tan said. "We just wanted to come by and see if you guys were still here. Everyone knows Cot, right? He helped me with my project and came with me to the banquet tonight."

Cos could see that Tan was trying not to break into a grin. "And? How was the banquet?"

"Well…" Tan paused, trying to delay her announcement. "I won! My design for the campus building was selected!"

Sin, Cos, and Csc jumped up, cheered, and congratulated Tan, while Sec glared at her, annoyed by those intruding green eyes again.

"Thanks," Tan said. "Cot contributed to my project too. As a graduate student in civil engineering, he did the plot survey."

"Well, while we're all here together at the tryst…" Sin paused while everyone laughed except for Sec, who glared stone faced. "Let's all sit down if everyone can stay a little longer. I know it's cold, but we haven't felt a drop of rain yet."

Csc returned to the bench where Sec was already sitting. Tan and Cot sat on the second bench, and Sin and Cos sat down together on the third bench. Tan expected to see Sin take Cos's hand, but Cos was surprised to see Cot take Tan's hand. She looked forward to talking with her about that development when they were back in their apartment. This location didn't seem like a tryst for Sec and Csc, as they both had their hands in their own jacket pockets and sat with as much space between them as the bench would allow.

Sin began in a sobering tone. "I think that while all six of us trigonometric functions are here in one place, we can talk with one another and come to an understanding of how much we need and depend on one another to make trigonometry work." He looked at each person while he talked. Even though Sec had his head down, Sin knew he was listening.

"You're saying that even though we have our individual properties and personalities," Tan said, "we still need to work in harmony with one another, or we become extremely limited. Not one of us functions stands alone in trigonometry. We all rely on one another."

"That's right," Sin said.

"I think that's what my professor was trying to tell Cot and me tonight at the banquet," Tan said.

"We all have a part to play," Cos said.

"Yes, we do. We've got to realize we've been working together all this time and didn't even know it," Sin said. "All six of us. Regardless of whether we've been getting along or not, we've been relating to one another according to our properties."

"If we are all equal, why do some of us feel so secondary?" Csc asked. Cot nodded his head in agreement, while Sec still sat in silence.

Tan said, "I think these feelings are self-imposed. We may be categorized at times as primary and secondary functions. Cot feels like the reciprocal of me, and he is, but does this make me better than he is? Do these categories make us primary functions better than secondary functions? I don't believe this. Reciprocal can mean mutual action or relationship. Sin is right. We all have an important part to play that no one else can, but we are dependent on one another. Even though I am a primary function, I have always felt a weakness like you, Sec, and Cot have, because I have asymptotes. They make me feel inconsistent and undefined at times, but if we look at asymptotes through a new perspective, they are actually strengths."

"Where have you found asymptotes to be strengths?" Cot asked.

"I've been doing some research. Business applications and optimization, to name a couple of areas," Tan said. "We should make friends with the calculus people to help us develop our potential even further and find more uses for our talents. Our weaknesses may actually be strengths in some applications if we view ourselves from a different perspective."

No one seemed to feel the cold anymore, as they were all absorbed by the conversation. The thunder retreated, and the clouds parted, allowing the crescent moon to reveal itself again. No one ever felt a drop of rain.

"I don't know," Sec said. "When would I ever work with Tan?" He looked across the cove straight into Tan's green eyes. His anger was dissipating but not completely resolved. His guard was still up as Tan looked directly back at him.

Sin thought for a second. "You've been working with Tan all this time and just haven't realized it. Have any of you heard of the Pythagorean identities?"

They all either shook their heads or remained still.

"All six of us have a component of ourselves in one of these identities. There are three of them:

1. $\sin^2 \theta + \cos^2 \theta = 1$
2. $\tan^2 \theta + 1 = \sec^2 \theta$
3. $1 + \cot^2 \theta = \csc^2 \theta$

"Divide the first one by cos all the way through, and you get the second one. Divide the first one by sin all the way through, and you get the third one."

"Csc, I had no idea I had been working with you," Cot said.

"Me neither." Csc smiled.

Sin continued. "Tan and Cot, do you realize how much you influence us with the tangent and cotangent identities?

$$1. \quad \tan\theta = \frac{\sin\theta}{\cos\theta}$$
$$2. \quad \cot\theta = \frac{\cos\theta}{\sin\theta}$$

"We could even relate these identities in terms of Csc and Sec if we wanted to. We would just substitute $\sin\theta$ for $1/\csc\theta$ and $\cos\theta$ for $1/\sec\theta$. We all relate to one another."

Looking around the cove, Cos observed, "Look how we have paired off into cofunctions; we are matching with our complements." Tan and Cot smiled at each other, and Sin squeezed Cos's hand. Sec and Csc looked at each other quickly and then looked away. The cozy feeling of being each other's tryst partner, cofunction, and complement didn't settle with either of them comfortably yet.

Csc started to laugh. She looked at Cos and Tan. "Remember when we looked up our complements that Unit Match predicted for us? They were right!"

$$1. \quad \sin\left(\frac{\pi}{2} - \theta\right) = \cos\theta$$
$$2. \quad \csc\left(\frac{\pi}{2} - \theta\right) = \sec\theta$$
$$3. \quad \tan\left(\frac{\pi}{2} - \theta\right) = \cot\theta$$

"Yes, they were," Cos said, laughing too.

"I thought it was so odd at the time that Unit Match predicted Cot as my complement," Tan said. "Now I can see why."

"Unit Match predicted a match between you and me?" Cot said to Tan. "Sweet. There was a time I thought

I would never have a chance with you. We're so reciprocal. I guess that's me feeling secondary again, which I shouldn't. We both have our strengths and weakness, but we complement and help each other."

"It's a good question to ask if everyone here feels OK talking about it," Sin said. "Csc, Sec, and Cot, can you talk with us about what it is that makes you feel secondary?"

After a moment of silence from the group, Cot leaned forward. "I liked and respected Tan from the moment I met her. I saw she was very principled and uncompromising. I wanted her to like me too, but I felt I couldn't match her principled mind-set and unrelenting high expectations. I felt secondary to her as well as reciprocal to her."

Everyone was listening intently, heads nodding, able to empathize. Even Sec, who continued to remain silent, seemed to agree.

Csc jumped in. "I've always felt secondary to Sin. He has a calm demeanor and shows depth and strength.

He's anchored, while I've always been blown every which way to get anyone to like me. He's also always been kind to me, even when I've done some despicable things I'm not proud of."

Sec's head lowered farther. He rubbed his forehead and made a choice not to have his explosive temper get the better of him, although he felt low and embarrassed for the first time with this group.

The five other trigonometric functions patiently sat for a few moments longer in silence to give Sec a chance to say something if he wanted to. The light breeze rustling the leaves in the trees was the only sound until Tan said, "It feels like we should have a campfire in the middle of this cove."

"I'll bring the hot dogs next time," Cot volunteered.

"I'll bring the marshmallows," Cos said.

Light laughter followed while Tan and Cot sat closer to each other to keep warm. Cos laid her head on Sin's shoulder. Sec and Csc still kept their distance.

## The Trigonometry Tryst

Finally, Csc looked at Sec and calmly suggested, "Do you have anything you'd like to share, Sec?"

Sec looked back at Csc, searching her eyes. He felt a sensation he wasn't used to. He had hurt a lot of people, resulting in his own embarrassment. He wasn't looked up to as he thought he was. This group thought he was a fool. He had called Csc a loser. *This group thinks I'm a loser.*

The group heard Sec take a deep breath before he said, "Cos, I think you look down on me. I'm not used to feeling secondary to anyone. I'm your reciprocal, but I'm a star athlete on campus, and you've never given me the time of day. I set up the Invitation to Meet at the $0°$ gate just for you. I wanted to take you out, and you just walked away. I had you come onto the basketball court with me at the end of our championship game. You walked away from me again that night also. I did it all for you."

"Except the Invitation to Meet was a fraud, wasn't it? You were an imposter. You led me to believe I was going to meet the person who represented the profile attached to the Invitation to Meet, but it wasn't your profile, was it?"

Sec cleared his throat. "No, it wasn't."

"It was Sin's profile," Cos said.

"Yes, it was," Sec said. "I knew you were attracted to Sin's profile, so I thought by using Sin's profile, I could at least lure you to the meeting."

"And you thought that was the best way to attract me to you?" Cos said.

Sec put his hands up. "All right, all right. I shouldn't have done it."

All eyes were on Sec as Sin asked, "Is there anything else you want to clear the air on? If you do, we'll help you work through it."

Running his hands through his hair, Sec sighed heavily and said, "I've used Csc shamefully. She's smart. I thought I could take advantage of her intellect combined with her low self-esteem to get my dirty work done. It worked until she discarded her low self-esteem. I see strength in her now that disgraces me. She's a different person. Transformed."

Csc and Sec looked at each other. Even in the dim light of the streetlamp, Sec saw light in Csc's intelligent hazel eyes. He began to feel a desire in himself. He wanted that light too.

"Sin, I'm responsible for tampering with your Invitation to Meet. I'm responsible for the failure of the electric bus the day Cos was riding it. I wanted to prevent the two of you from spending time together. I'm appalled at myself for treating you all the way I did at the fraternity house the night of the championship game. I'm sorry for it all. I don't know how long it will take for all of you to forgive me, if you ever do, or for me to improve myself, but I know I feel a power in this group of people right here, a power I haven't sensed before. I'd like to be part of it."

"Keep being a championship athlete, Sec," Sin said. "See how you can use your popularity and influence for the better instead. By the way, congratulations on your winning shot."

Sec stood up, walked over to Sin, and shook his hand. The rest of the group stood up and closed the circle, shaking one another's hands. When Sec held out his hand to Tan, he said, "Tan, your eyes aren't so threatening to me anymore." Cot smiled in understanding.

Turning to face Csc, Sec said, "You taught me something. I never would have realized I've been going down the path of failure if it weren't for you. Can we start again? Be my study partner at the café?" He softly tugged one of her blond curls and let it spring back to its resting position.

"We'll see," Csc said softly but directly. "I won't make that decision now."

"Fair enough," Sec said.

The sky began to clear as the six trigonometric functions looked up at the twinkling stars.

"Think of how our skills have been used for centuries to help explorers and travelers navigate the globe by referencing the stars," Sin said. "We can solve anything together, working in harmony and keeping our relationships intact. We need to 'hear' one another, as in 'The Telegraph.' Trigonometry is powerful. We are powerful."

The light from the faces of the six trigonometric functions looking up at the stars reflected this power.

## ACKNOWLEDGEMENTS

● ● ●

I WOULD LIKE TO THANK Mary Lynn Bailey for her editing advice and my family for their encouragement on this project.

## ABOUT THE AUTHOR

● ● ●

J. A. BAILEY HAS A bachelor of science in metallurgical engineering and a master of business administration. Bailey loves mathematics, reading, and traveling, as well as spending time with her nieces and nephews.

Bailey has dedicated a portion of this book's proceeds to a scholarship fund that aids qualified students in their educational pursuits.

www.ingramcontent.com/pod-product-compliance
Lightning Source LLC
Chambersburg PA
CBHW071411180526
45170CB00001B/53